Textbook on Gender and Women Empowerment

About Authors

Karthik Dharamkar has graduated in B.Sc. Agriculture in the year 2012 and M.Sc. Agricultural Extension in the year 2014 from College of Agriculture, Rajendranagar of ANGRAU, Guntur, Andhra Pradesh. He is currently pursuing his Ph.D. in Agricultural Extension Education at ICAR-National Dairy Research Institute (NDRI), Karnal, Haryana.

Hema Sarat Chandra N has graduated in B.Sc. Agriculture from Agricultural college-Naira of ANGRAU-Guntur, Andhra Pradesh in 2011. Thereafter, he has attendedto Banaras Hindu University (BHU, Varanasi) and completed his M.Sc. Agriculture in Extension Education in 2013. By securing UGC JRF he has done his Ph.D. in Agricultural Extension education from ICAR-National Dairy Research Institute (NDRI), Karnal, Haryana in 2016. He has evaluated the functioning of Rythu bazaars (farmers market) in Visakhapatnam, Andhra Pradesh as a part to complete his M.Sc. research. He has worked on Agricultural Technology Management Agency (ATMA), Andhra Pradesh for his Ph.D. research and submitted thesis as Dynamics of coordination among different agencies with ATMA in Andhra Pradesh. He has developed ATMA coordination index (ACI) and Group dynamics index (GDI) to measure the coordination of ATMA and group dynamics of farmer interest groups under ATMA. He has published research articles in national as well as international journals in agriculture and rural development sectors.

Gajanand Palve has graduated in B.Sc. Agriculture from Agricultural college- Jagitial of ANGRAU in 2014. Thereafter, he has attended to Bidhan Chandra KrishiVishwavidyalaya (BCKV), Mohanpur, West-Bengal and completed his M.Sc. Agriculture in Agricultural Extension in 2016. He has secured ICAR-SRF in Agricultural Extension and presently pursuing Ph.D. in Agricultural Extension education from ICAR-National Dairy Research Institute (NDRI), Karnal, Haryana. He has specialised in studying technology adoption behaviour of farmers.

Niketha L She has graduated in B.Sc. Agriculture from College of Agriculture, Shimoga while M.Sc. Agricultural Extension Education from College of Agriculture, Pune and her Ph.D. in Agricultural Extension Education at ICAR-National Dairy Research Institute (NDRI), Karnal, Haryana.

Textbook on Gender and Women Empowerment

Karthik Dharamkar
Hema Sarat Chandra N
Gajanand Palve
Niketha L

2018

Daya Publishing House®

A Division of

Astral International Pvt. Ltd.

New Delhi – 110 002

Published by : **Daya Publishing House®**
A Division of
Astral International Pvt. Ltd.
– ISO 9001:2015 Certified Company –
4736/23, Ansari Road, Darya Ganj
New Delhi-110 002
Ph. 011-43549197, 23278134
E-mail: info@astralint.com
Website: www.astralint.com

Digitally Printed at : **Replika Press Pvt. Ltd.**

Preface

The Millennium Declaration identified Gender Equality and Women's Empowerment (GEWE) as one of eight Millennium Development Goals and stated that it was an effective means to combat poverty, hunger and disease, as well as to stimulate development that is truly sustainable. Millennium Development Goal 3 (MDG 3) was established to "Eliminate gender disparity in primary and secondary education, preferably by 2005, and at all levels of education no later than 2015." The MDG Summit 2010 called for further action to ensure gender parity in education, health, economic opportunities and decision-making through gender mainstreaming in development policymaking. An important route to achieving gender equality is by empowering of women through education, employment and political representation, as well as by ensuring women's access to reproductive health services. Another fundamental step towards the realization of gender equality is to eradicate all forms violence against women. As consideration of these facts this book covers all gender and women empowerment related concepts and theories in a well designed manner.

This book is prepared as per the ICAR syllabus by covering all the gender and women empowerment related concepts and terminologies. This book is very useful for UG, PG, and Ph.D. students those who has syllabus related to gender and women empowerment. This book is very useful for competitive exams like UG, PG entrance examinations of ICAR-SAU'S, JRF, SRF, and ICAR-NET. Constructive suggestions of readers are invited for further improvement of this book.

Karthik Dharamkar

Hema Sarat Chandra N

Gajanand. Palve

Niketha L

Contents

Chapter 1

Gender Concepts and Terminologies

Δ Gender

- "Gender" refers to the socially constructed roles and responsibilities of women and men, in a given culture or location.
- People often use the word "Gender" as a synonym for "Sex." Sex, however, refers to a biological characteristic that makes someone female or someone male. We also misuse the word Gender as a synonym for "women" or "female
- Gender is a social construct that defines what it means to be a man or a woman in a given society.
- Gender often prescribes roles for men and women, dealing with division of labor, responsibilities and rights.
- These roles vary by culture and may change over time. In societies, gender structures have the potential to produce institutionalized inequalities between men and women
- The word 'gender' originated from the French word 'Genil' meaning how society and culture define female and male.
- The word gender is also not new to us, we all learnt about this in our first lesson of grammar to classify masculine gender and feminine gender. But in Sociology, it is used as a conceptual category and a very specific meaning is given to.
- Thus, Gender as a 'term' explains the way society constructs the differences between women and men and girls and boys.
- Sex describes the biological differences between men and women, which are universal and determined at birth.
- Gender refers to the roles and responsibilities of men and women that are created in our families, our societies and our cultures.

- Gender is not permanent. Gender changes from time to time with increasing rapidity due to change in cultural, economic development, technology, constitutional and legal-framework.
- A Basic Distinction-Sex and Gender: Sex refers to the biological characteristics between men and women, which are universal and do not change. These sets of biological characteristics are not mutually exclusive as there are individuals who possess both, but these characteristics tend to differentiate humans as males and females. As race, class, age, ethnic group, etc the notion of gender needs to be understood clearly as a cross-cutting socio-cultural variable. Gender refers to social attributes that are learned or acquired during socialisation as a member of a given community. Gender is therefore an acquired identity. Because these attributes are learned behaviours, they are context/ time-specific and changeable (with increasing rapidity as the rate of technological change intensifies), and vary across cultures. Gender therefore refers to the socially given attributes, roles, activities, responsibilities and needs connected to being men (masculine) and women (feminine) in a given society at a given time, and as a member of a specific community within that society. Women and men's gender identity determines how they are perceived and how they are expected to think and act as men and women. Gender determines what is expected, allowed and valued in a woman or a man in a given context. In most societies there are differences and inequalities between women and men in responsibilities assigned, activities undertaken, access to and control over resources, as well as decision-making opportunities. Other important criteria for socio-cultural analysis include class, race, poverty level, ethnic group and age.
- It is worth noting that when the word "gender" was first used in this way, to signify social rather than grammatical difference, at the Institute for Development Studies at the University of Brighton, in the mid-1970s, the intention was not so much as to distinguish men from women, 2 but to find an analytic tool to disaggregate the category "women". In other words, the intention was to refine analysis of the differing impacts of development on different groups of women, as well as differences between women and men.

Δ Masculinity and femininity

- Patterns of differences by gender is seen when the character is either masculine or feminine.
- Masculinity and femininity are concepts which signify the social outcomes of being male or female including the traits and characteristics which describe men and women and often give men advantage over women.
- Hence, the normal male has a preponderance of masculinity and therefore, gender is the amount of masculinity or femininity found in a person
- Some cross-cultural elements, such as aggression, strength, and assertiveness have traditionally been considered male characteristics.

- The socially and historically constructed male characteristics need to be seen in their specific historical, cultural, and social context.

Δ Patriarchy

As a concept to refer to the social system of masculine domination over women. Through such differential treatment, women are denied access to resources of the society and to positions of power and authority both in the family and in the community.

Δ Women in Patriarchy

Because patriarchy is a system, women are also patriarchal. They co-operate in perpetuating patriarchy and are rewarded by the system for doing so. Thus the fight is not with all men, but it is with those men and women who perpetuate patriarchy.

Δ Women in Development (WID)

WID subscribes to the assumptions of modernization theory. Its programmes generally stress western values and target individuals as the catalysts for social change. Modernization theory identifies traditional societies as male-dominated and authoritarian compared to modern societies which are democratic and egalitarian. It usually seeks to integrate women into development by making more resources available to women. However, these efforts led to increase in women's work load, reinforced inequalities, and widened the gap between men and women.

Δ Women and Development (WAD)

It emerged from a critique of the modernization theory. The theoretical base of WAD is dependency theory and focuses on relationship between women and development process and examines the nature of integration. It is concerned with women's productive role and assumes that once organizational structures become more equitable, women's position would also improve.

Δ Gender and Development (GAD)

The gender and development seeks to base interventions on the analysis of men's and women's roles. This approach was developed in the 1980s. It questions the basis of assigning specific gender roles. Recognizes that patriarchy operates within and across classes both inside and outside the home and oppresses women.

Δ Gender Sensitization

Gender sensitization is the process of changing the stereotype mindset of men and women, a mindset that strongly believes that men and women are 'unequal entities'. Its goal is essentially to create a value system in society that accords explicit and spontaneous recognition to the contribution of women in socio-economic development, and respects their wisdom; a system that makes women sensible and courageous enough to recognize their own contribution and make them feel proud of.

Gender sensitization should seek to change not only the impression of men towards women i.e. the way men think of and treat women, but also the attitude of women i.e. the way women think of men and of themselves and their behaviour in this context.

Δ Gender bias

Perception that both sexes are not equal and do not have similar rights to resources.

Δ Gender Identity

Person's private, subjective sense of their own sex.

Δ Gender analysis

Gender analysis is the first and most critical step forward towards gender-responsive planning and programming. It involves the collection and analysis of sex disaggregated information. It examines the differences, commonalties and interactions between women and men. Gender analysis examines women and men specific activities, conditions, needs, access to and control over resources, access to development benefits and decision-making. It directs "WHO DOES WHAT" and provides us with precise information on gender perspective that is required for gender sensitization. The answers are based on facts regarding the patterns in the lives of women and men.

It refers to the study of differences in conditions, needs, participation rates, access to resources, control of assets, decision-making powers etc. between women and men in their assigned gender roles. It is the methodology for collecting and processing information.

Gender Analysis: Is the methodology for collecting and processing information about gender.

Gender analysis helps to understand the relationships between men and women. Assessing the relationship makes it possible to determine men's and women's constraints and opportunities within the farming system. The analysis answers the following questions:

- Division of labour: who within the household carries out which agricultural tasks (What do men and women do?)
- Who has access to and control over resources and services?
- How resources distribute and who makes the decisions (what decisions do men and women make in the family/community)?
- What are the reasons behind these differences in gender?

Δ Why gender analysis is done?

- To better understand the opportunities/problems in the community and plan interventions which are beneficial to both women and men.
- To expose the barriers to women's full participation and economic development
- To make decisions & implement the project/programme that promotes gender equity.
- It helps to find the best strategies and solutions to address the different needs and dynamics of men and women living in poverty.

Δ Gender analysis can be done using:

- Formal interviews and surveys, mapping
- Household interviews and focus group discussions (FGDs)

One of the approaches to conduct gender analysis is using household interview. By conducting a household interview a lot of information (qualitative/quantitative) can be obtained that can give better understanding about the issue. For example, in semi-structured interviews general questions or topics are initially identified and become basis for more specific questions during the interview allowing both the interviewer and the person being interviewed discuss on the issues. The other approach is focus group discussion. In this case small groups (usually 5-10 people) are formed for an open discussion assisted by a facilitator/moderator. Unlike individual interviews focus group discussion provides an added dimension of the interactions among members.

The Main Goals of Gender Analysis

- To better understand our communities
- To promote gender equality through our work
- To expose the barriers of women's full participation and economic development.
- To help us find the best strategies and solutions and to address the deferent needs and dynamics of men and women living in a poverty.

When Should We Use Gender Analysis

- Evolution of the policy with principles
- Planning
- Programming
- Budgeting
- Implementation
- Monitoring and
- Evaluation

The Aim of Using Gender Analysis in Agriculture is To:
- Improve overall project performance
- Overcome gender-based barriers
- Promote equal opportunities
- Increase both men's and women's participation.
- Ensure that new technologies will not have an adverse impact on women

Gender analysis is an effort to understand how gender issues relate to development processes, through the application of a set of questions and tools that are to be integrated in all steps of the project. Therefore, it is imperative to ask how a particular activity, decisions or plan will affect women differently from men, as the analysis is based on the fact that women and men play different roles in society, connected with different problems, different needs and priorities. It is for this reason that gender analysis must be applied at all stages of the development process (MOWA, 2009).

To conduct a Gender Analysis, a core set of issues should be addressed. These are:

Women's and men's roles.	Who does what, with what resources? Paying particular attention to variations within
	subgroups of women and men (eg. Elderly women, adolescent girls, men from urban areas etc). Typically, women perform three kinds of roles:
	-Productive roles (paid or not);
	-Reproductive roles (sustaining family living conditions and basic needs-usually unpaid work) and
	-Community role.
Factors that shape gender roles and the gender division of work	Depending on the circumstances, traditions and institutions that shape gender roles represent constraints and / or opportunities for women and men. Understanding to what extent, and when, they are critical to designing the programs and projects suitable to the community

Access to and control over resources and opportunities, and their systems of distribution	Not all men and women have the same access to and control over resources and opportunities. Understanding the mechanisms and rules by which the resources and benefits are distributed is important to assess the situation of women vis-à-vis men (and vice versa) and determine the most effective entry points for action
Access to and participation in decision making processes	Who decides? How are decisions taken concerning women's and men's lives and those of their families? Are women and men equally represented or given an opportunity to influence such processes?
Men's and women's practical and strategic needs and interests	Given their respective roles, who needs what for what purpose?

Δ Gender analysis tools/frameworks

Gender analysis tools/frameworks are approaches used to generate data & information during gender analysis. They answer questions such as: who does what, who has what, who needs what and what should be done to close the gaps between what women and men need. In order to conduct gender analysis various tools have been developed among which them are presented below:

1. Harvard Analytical Framework/gender roles framework
2. Moser (triple roles) Framework-caroline moser
3. Levy (web of institutionalisation) Framework
4. Gender Analysis Matrix (GAM)
5. Equality and Empowerment Framework (Longwe)
6. Capacities and Vulnerabilities Framework
7. People Oriented Framework (POF)
8. Social Relations Framework (SRF)-neila kabeer
9. SEAGA Approach (The Socio-economic and Gender Analysis)
10. A useful framework is developed by DFID (department for international development, U.K). It is designed to guide gender analysis at the primary stakeholder/community level. Its use/emphasis should be adapted to the particular situation and sector.

Δ   **The Harvard Analytical Framework, also known as the Gender Roles Framework:**

The Harvard Framework was one of the earliest gender frameworks, having been developed by the Harvard Institute of International Development in collaboration with USAID's Women In Development Office in the early 1980s. It approaches gender analysis as means of increasing the efficiency of development programme; by mapping gender divisions of labour and resources it aims to demonstrate the important role that women play in economic productivity.

The Objective of this framework is

- To demonstrate that there is an economic rationale for investing in women as well as men.
- To assist planners design more efficient projects.
- To emphasise importance of good information as basis for efficient/ effective projects.
- To map the work of women and men in the community and highlight differences.

Features

The framework consists of a matrix for collecting data at the micro (community and household) level. It has four interrelated components:

- o   the activity profile, which answers the question, "who does what?", including gender, age, time spent and location of the activity
- o   the access and control profile, which identifies the resources used to carry out the work identified in the activity profile, and access to and control over their use, by gender
- o   the analysis of influencing factors, which charts factors that influence gender differences in the above two profiles
- o   the project cycle analysis, which examines a project or intervention in light of gender-disaggregated information

The framework also contains a series of checklists consisting of key questions to ask at each stage of the project cycle: identification, design, implementation, and evaluation.

Benefits of the framework

- Best suited for project planning, rather than programme or policy planning
- As a gender-neutral entry point when raising gender issues with constituents resistant to considering gender relations and power dynamics
- For baseline data collection

- In conjunction with Moser's framework, to draw in the idea of strategic gender needs

Strengths and Weakness of Harvard framework

Strengths	Limitations
• Practical and easy to apply	• Focus on material resources rather than strategic gender needs (e.g., increasing women's access to decision making)
• Helps collect and organize information about gendered division of work	
• Helps make women's work visible	• Tends to oversimplify and homogenize inequalities (e.g., differences between mothers-in-law and daughters-in-law)
• Distinguishes between resource access and control	
• Adaptable to a variety of settings and situations	• Ignores underlying inequalities, such as class, race, or ethnicity (e.g., poor women and wealthy women will not necessarily perform the same kind of agricultural or household work)
• Non-threatening inquiry about women or men's activities because it relies on "facts"	

Activity Profile Tool (example)

	Women	Men	Girls	Boys
Productive Activities				
Agriculture				
° activity 1				
° activity 2				
Income generating:				
° activity 1				
° activity 2				
Employment				
Reproductive Activities				
Water				
Fuel				
Cooking				
Childcare				
Health				
Cleaning				
Repair				

Community Involvement	
Meetings	
Wedding	
Helping others	
in time of need	
Labor on communal	
projects	

Access & Control Profile Tool (example)

	Access		Control	
	Women	**Men**	**Girls**	**Boys**
Resources				
Land				
Water				
Seed				
Labor				
Extension				
Etc.				
Benefits				
Assets				
Income				
Education				
Political				
power/				
prestige				
Etc.				

Influencing Factors Tool (example)

	Women	**Men**	**Girls**	**Boys**
Factors				
-Community norms				
-Social hierarchy				
-Institutional				
-Economic				
-Political				

Δ The framework consists of three interrelated components:

An activities profile, listing the tasks men and women do (disaggregated by age, ethnicity, class etc) and where and when these tasks are performed. Activities are grouped into the three areas of productive activities, reproductive or household activities and social/political/religious activities

- An access and control profile, listing the resources needed to carry out the above activities and the benefits that result from them. Resources include political and social resources such as education as well as material and economic resources such as land or capital.

- An influencing factors profile, which identifies the factors that affect the division of labour and resources. These may include general economic conditions (such as poverty levels, income distribution, infrastructure etc), institutional structures (such as government bureaucracies), demographic factors, community norms and social hierarchy (such as family/community power structure and religious beliefs etc), legal parameters, political events (internal and external), national attitude to refugees and attitude of refugees to development/assistance workers.

The Harvard Framework has been successful in that it assists in making women's labour visible to those designing projects and programmes. Moreover, it illuminates the inequality women often face in accessing, and especially controlling, resources. The distinction made between access and control shows gender inequality particularly starkly. By looking at control of resources the way is also paved for discussions of power relations, albeit not the original intention of this framework.

However, the simple structure of the Harvard Framework tends to simplify the complex social issues involved in gender and development and does not allow room for considering the ways in which men and women's roles and activities cooperate and interrelate. Some have criticised the approach of Harvard Framework for being too focused on efficiency and material resources and insufficiently aware of the social dynamics of gender inequality, thus not providing much guidance on how to change socially entrenched gender inequality.

Moreover, the Harvard Framework relies on detailed knowledge of a situation for it to be effective and this may not be available at early planning stages of a project. It is also difficult to use over a large region where circumstances may differ considerably. Furthermore, it has been criticized for not being participatory

Δ The Moser Gender Planning Framework

This framework, developed by Caroline Moser, links the examination of women's roles to the larger development planning process. The framework introduces the idea of women's "three roles" in production, reproduction, and community management, and the implication that these roles have for women's participation in the development process. In making these links, both between women and the community, and between gender planning and development planning more broadly, Moser's framework encompasses both the technical and political aspects of gender integration into development.

The Moser framework follows the Gender and Development approach in emphasizing the importance of gender relations. As with the WID-based Harvard Analytical Framework, it includes collection of quantitative empirical facts. Going further, it investigates the reasons and processes that lead to conventions of access and control. The Moser Framework includes gender roles identification, gender needs assessment, disaggregating control of resources and decision making within the household, planning for balancing the triple role, distinguishing between different aims in interventions and involving women and gender-aware organizations in planning. The framework acknowledges a political element to gender planning, and assumes that the process will have to deal with conflicts.

The Objective of this Framework is

Focus on strategic gender needs and concentrates on gender inequalities and how to address these at programme and policy level.

The framework is composed of several components (or tools). The framework includes six tools:

1. Gender roles identification. This tool maps gendered division of labour within the household by examining the activities all women and men, girls and boys, over a twenty-four hour period. Moser identifies a triple role for low income women: productive work (income generation and the production of goods and services for the consumption of the household), reproductive work (bearing and rearing children, domestic tasks that maintain household members such as cooking, cleaning, and caring for the sick and elderly) and community roles (provisioning and maintaining resources used by everyone). All of these roles for women are often underappreciated because they tend to be non-paid and because productive and community roles closely relate to reproductive roles which are themselves undervalued.

2. Gender needs assessment. Here Moser uses Maxine Molyneux's 1984 distinction between practical gender needs and strategic gender interests/ needs. Practical gender needs are those identified by women in their daily lives and which would improve living conditions such as safe access to water, better maternal health care or a cash income. However, addressing only practical Gender needs is unlikely to address gender inequality. Strategic gender interests/needs, on the other hand, do just that. These may include equal pay, legal rights, changing gendered division of labour and ending sexual and gender based violence (SGBV). This is not a clear cut distinction in many cases but a continuum of needs from practical to strategic.

3. Disaggregated control of resources and decision-making within a household. This identifies who controls the household's resources, who

makes decisions about those resources and how those decisions are made.

4. Balancing of roles. This tool considers women's workload from their three roles and how they balance their responsibilities. It endeavours to ensure that any intervention will not add to that workload.

5. WID/GAD policy matrix. Assessing the extent to which programmes fit into WID or GAD policy approaches.

6. Involving women, gender aware organisations and planners in planning. This tool tries to ensure that women participate in this process and that it addresses the needs that they themselves experience as opposed to needs that project planners may perceive them as having.

This framework is successful at making all work visible to planners and the interrelationships of different forms of work. Moreover, it helps address issues of social and political inequality through the exploration of strategic gender needs, as well as recognising the likelihood of institutional and political resistance to changes in gender roles.

However, Moser's framework has been criticised for failing to address differences of age, class, ethnicity etc, and therefore treating women as a homogenous group. Moreover, this framework does not take into account men's gendered need or the ways in which they may intersect with women's.

Best Suited

- For planning at all levels from policies to projects.
- In conjunction with the Harvard Framework

Strengths and Limitations

Strengths	Limitations
• Assumes planning exists to challenge unequal gender relations and support women's empowerment.	• Framework does not mention other inequalities like class, race and ethnicity.
• Makes ALL work visible through concept of triple roles.	• Framework is static and does not examine change over time.
• Alerts planners to interrelationships of triple roles.	• Looks at separate, rather than inter-related activities of women and men.
• Recognises institutional and political resistance to transforming gender relations.	• Strict division of practical and strategic needs not always helpful in practice.
• Distinguishes between practical gender needs (those that relate to women's daily life) and strategic gender needs (those that potentially transform the current situation)	• Strategic needs of men not addressed.

Using this Framework in Agricultural Extension and Advisory Services

1. With the participation of men and women together or in separate groups, identify and map out all the activities of men and women in the household over a 24-hour period using the "triple roles" framework. Include girls and boys in the mapping exercise to obtain a fuller picture.

2. Clarify the needs of men and women by distinguishing between their practical needs (e.g., is there food to cook for dinner?) and strategic needs (e.g., access of women to decision-making or women's control of their time).

3. Discuss results of 1 and 2 with the group.

4. Use this information to assess how men and women manage their various roles and deciding on the timing of extension programs that are suitable to women's available time and does not create additional burden on women's responsibilities in the household.

5. Identify the type of extension interventions that you can tailor to men and women's roles and their needs.

6. Involve men and women during the planning process to design and offer gender-effective extension programs.

Δ The Gender Analysis Matrix (GAM)

The Gender Analysis Matrix (GAM), developed by Rani Parker, is a tool that uses participatory approaches to identify how a particular agricultural production, processing or marketing practice impacts men and women differently in the community. Its community-focused approach enables participants to analyze differences between men and women's culturally-defined practices in the community, and challenge their assumptions about gender roles.

For example, a community could evaluate the labor practices of men and women, and assess how they impact men and women's wellbeing, time use or earning potential. If they identify a particular activity that discriminates against women, the community may decide to stop such a practice and create a more equitable environment.

The benefits of using GAM are based on its community-based techniques to collect and analyze information on the spot. The targeted community self-identifies problems and finds equitable solutions prompting transformative change. The use of participatory approach, where a group of men or women, or mixed group of men and women identify and propose culturally-validated changes at the community level.

Basics of GAM

To ensure effective gender analysis process with the use of GAM, it is important to understand that:

- Participating men and women possess all necessary information about themselves and their community.
- Application of GAM doesn't require the community to have gender expertise, but extension agents facilitating the process should possess an understanding of gender analysis.
- Application of GAM facilitates transformative change in the community as the community self-identifies problems and consensually validates its own solutions.

Parker (1993) describes GAM using What, Why, Who, When and How

- WHAT: A tool for gender analysis of development (program/ project) at community level.
- WHY: To determine differing impacts of interventions on women & men.
- WHO: Done by a group within the community; includes women & men in equal number
- WHEN:
  o At the planning stage --> determine potential gender effects
  o At the design stage --> gender considerations may change the design
  o During monitoring stage --> address broader program impacts

The framework

The GAM is based on the following principles:

- All requisite knowledge for gender analysis exists among the people whose lives are the subject of the analysis.
- Gender analysis does not require the technical expertise of those outside the community, except as facilitators.
- Gender analysis cannot promote transformation unless it is carried out by the people being analysed.

GAM Matrix

Intervention	Labor	Time	Resources	Culture
Men				
Women				
Household				
Community				

GAM Tool 1: Level of Analysis

GAM allows analysis of an intervention at four levels: men, women, households, and community. The levels of analysis appear vertically on the matrix:

- Men: Represent men of all ages who are in the target group or all men in the community.
- Women: Represent women of all ages who are in the target group or all women in the community.
- Household: Represents all women, men, and children living under one roof.
- Community: Represents everyone in the community.

GAM Tool 2: Impact Analysis

GAM examines impact on four areas, which appear horizontally on the matrix:

- Labor: Captures changes in tasks (do women take over men's tasks in the field), the level of skill (formal education, training) required, the number of people involved in this activity and the demand for additional labor.
- Time: Captures changes in time requirements to complete specific tasks.
- Resources: Captures changes in access to resources (income, land, extension information) and the extent of control over resources (increase or decrease) as a result of an intervention.
- Socio-cultural factors: Captures changes in gender roles or status as a result of an intervention.

Strengths and Limitations

Strengths	Limitations
• Designed specifically for community-based development workers	• Needs a good facilitator
• Simple and systematic uses familiar categories and concepts	• Some factors can get lost because categories have many aspects
• Transformatory as well as technical	• Requires careful repetition in order to consider change over time
• Fosters 'bottom-up' analysis through community participation	• Does not seek out the most vulnerable community members
• Considers gender relations between women and men, as well as examining what each category experiences separately	• Excludes macro-and institutional analysis
• Levels of analysis can be added to in order to suit particularly interventions	• Difficulties defining a community
• Includes intangible resources	• Subordination is often no explicit
• Can be used to capture changes over time	• Risk of misleading outcomes due to power relations between funders and community members

- Helps anticipate resistance, and encourages consideration of what support should be offered for those at risk
- Includes men as gendered beings, so can be used in interventions that target men

- Can be used for participatory impact assessment
- Quick data gathering

Δ The Women's Empowerment Framework (WEP)

Sara Hlupekile Longwe developed the Women's Empowerment Framework (sometimes referred to as the Longwe method), arguing that women's inequality is a result of oppression and exploitation and therefore development means overcoming all forms of gender inequality. Poverty will only be addressed through women's empowerment.

Objective of the women's empowerment framework is

To achieve women's empowerment by enabling women to achieve equal control over factors of production and participate equally in the development process

The WEP Framework consists of five levels of empowerment that women can achieve and the first tool assesses to what level a project is operating. The levels are hierarchical and therefore projects that focus on the higher levels are more likely to deliver women's empowerment than those only operating at the lower levels. These five levels are:

1. Welfare: Equal material wealth (income, food supply, health care) for men and women.
2. Access: Equal access for women and men to the factors of production (land, credit, labour, training, marketing facilities, and public services) and equal opportunities in life. Usually this is achieved, at least in part, by abolishing discriminatory laws.
3. Conscientisation: Both genders understand the concept of gender and how it affects lives. The gendered division of labour should be fair and equal and have the approval of both men and women.
4. Participation: Women participate equally with men at all levels, including in development projects.
5. Control: Women and men reach a fair balance of control over resources. This is achieved through women's conscientisation and mobilisation.

The second tool involves an assessment as to the level of recognition that women's equality has received in the design of a project. There are three levels of recognition:

1. Negative. No reference to women's issues (defined by Longwe as anything relating to equality with men) in the project objectives. The project will probably have a negative impact on women by worsening inequality.
2. Neutral. Although women's issues are included the project is unlikely to actually improve equality of women.
3. Positive. An active interest is shown in women's issues and with improving women's position relative to men.

Framework distinguishes between women's issues and concerns as well as identifying three levels of recognition of women's issues in project design.

Longwe also distinguishes between

Women's issues	which pertain to equality with men in any social or economic role and involving any of the levels of equality
Women's concerns	which pertain to women's traditional and subordinate, sex-stereotyped gender roles

The women's empowerment framework identifies three levels of recognition of women's issues in project design:

Negative level	where project objectives are silent about women's issues. Experience suggests that women are likely to be left worse off by such a project
Neutral level	where the project objectives recognise women's issues but concern remains neutral or conservative, merely ensuring that women are not left worse off than before
Positive level	where project objectives are positively concerned with women's issues and with improving the position of women relative to men

The framework can be used to produce profiles as below:

Levels of Recognition	Levels of Equality		
	Negative	**Neutral**	**Positive**
Control			
Participation			
Conscientisation			
Access			
Welfare			

Best Suited

- Useful across micro (project) and macro (country strategy) levels of analysis.
- Useful where focus is specifically on empowerment of women

Strengths and Limitations

Strengths	Limitations
• Framework can be used to prepare profiles of levels of recognition as well as profiles of analysis of levels of equality across sectors.	• Assumption that a level of equality is strictly hierarchical is questionable.
• Develops notion of practical and strategic gender needs into progressive hierarchy.	• Framework profiles are static and do not take account of changes over time.
• Articulates empowerment as essential element of development.	• Focus on gender equality only takes no account of interrelationships between rights and responsibilities.
• Enables assessment of interventions based on grounds of empowerment.	• Ignores other forms of inequality.
• Has a strong political perspective-aims to change attitudes.	

The WEP framework is useful in that it can be used for planning, evaluation and monitoring, or training, and is therefore a versatile tool.

However, Longwe's concepts of power relations are simplistic, looking at men and women's relationship only in terms of equality (and therefore excluding consideration of systems of rights, claims and responsibilities between men and women, as well as ways in which they cooperate and support one another) and viewing women as one, homogenous group. Moreover, the hierarchical levels of empowerment falsely suggest that this is a linear and clear-cut process.

Δ The Social Relations Approach Framework (SRA)

The SRA was developed by Naila Kabeer at the Institute of Development Studies, University of Sussex, UK. Instead of developing specific and technical tools, it is based on 5 key concepts:

1. Development is increasing human well-being, key elements of which are survival, security and autonomy. Economic productivity is insufficient as a goal for development.

2.   Social relations determine rights, responsibilities, claims and roles. Kabeer understands social relations in terms of how groups are positioned in relation to resources. Social relations are constantly changing.

3.   Institutions produce inequality, including gender inequality. There are four levels of institutions: state, market, community, family. Kabeer defines an institution as a framework of rules for achieving particular economic or social goals and organisations as the specific structural forms that institutions take. Institutions can be considered according to their key aspects: rules (how things get done), activities (what is done), resources (what is used/produced), people (who is in/ out, who does what) and power (who decides and whose interests are served).

4.   Institutions operate according to different gender policies. Kabeer identifies 3 categories:

   •   Gender-blind policies which fail to distinguish between men and women and therefore are likely to be biased in favour of existing gender relations and to exclude women.

   •   Gender-aware policies which are divided into gender-neutral policies (use knowledge of gender differences in context to meet the practical needs of men and women without disturbing gender relations) and gender specific policies (use the same knowledge to meet the needs of men or women without addressing division of resources and responsibilities of each gender)

   •   Gender-redistributive policies which aim to transform the existing distribution of resources and responsibilities.

5.   Analyses of a situation need to examine where inequality is caused by immediate, underlying and/or structural factors and address it appropriately in their intervention.

The SRA provides a dynamic picture of the broader processes of poverty and inequality that puts gender right at the centre of development. It allows links to be made between what is going on at the grassroots level and at the national level. However, it has been criticised for being too complex for easy institutional use and for minimising agency of beneficiaries by focusing on institutions which sideline questions of individuals' abilities to affect change.

   •   Gender roles are those behaviours, tasks and responsibilities that a society considers appropriate for men, women, boys and girls

In some rural societies, commercial agricultural production is mainly a male responsibility. Men usually prepare land, irrigate crops, and harvest and transport produce to market. They own and trade large animals such as cattle, and are responsible for cutting, hauling and selling timber from forests.

Women and girls play an important, largely unpaid, role in generating family income, by providing labour for planting, weeding, harvesting and threshing crops, and processing produce for sale. Usually they are responsible for taking care of smaller animals.

In most societies rural women have also the primary responsibility for maintaining the household. They raise children, grow and prepare food, manage poultry, and collect fuel wood and water.

These gender roles can vary considerably depending on the geographical area, culture and other factors.

- Gender roles are affected by
    I.   Age
    II. Class
    III. Race
    IV. Ethnicity
    V.   Religion
    VI. Geographical, Economic & political environment

Δ Productive Work/roles:

Involves the production of goods and services for consumption and trade, (farming, fishing, employment and self-employment.)

Δ Reproductive Work/roles:

Involves the care and maintenance of the household and its members including bearing and caring for children, food preparation, water and fuel collection, shopping, housekeeping and family health care.

Δ Community Work/roles:

Involves the collective organization of social events and services, ceremonies and celebrations, community improvement activities, participation in groups and organisation , local political activities and so on. This type of work is seldom considered in economic analyses of communities.

Δ Community Politics Role:

Activities undertaken primarily by men at the community level, organizing at the formal political level, often within the framework of national politics. This work is usually undertaken by men and may be paid directly or result in increased power and status.

Δ Access to productive resources

It refers to right and opportunity of men and women to use the resources as per one's need to carryout his/her activities. .

Δ Control over productive resources

It refers to the rights and power of men and women to decide on the use and destination of the resources.

Δ Gender audit

The analysis and evaluation of policies, programmes and institutions in terms of how well they apply gender-related criteria.

Δ Gender budgeting

Gender-based assessment of budgets, incorporating a gender perspective at all levels of the budgetary process and restructuring revenues in order to promote gender equality.

Δ Gender equality

Gender equality means that all human beings are free to develop their personal abilities and make choices without the limitations set by strict gender roles. Different behaviour, aspirations, and needs of women and men are considered, valued and favored equally.

Δ Gender equality as a goal

The aim of UK international development policy is to contribute to the elimination of world poverty. The empowerment of women and the promotion of gender equality is one of the eight internationally agreed development goals designed to achieve this.

Gender equality is given such high priority because:

- **Gender equality is essential to poverty elimination.**
  There is a growing and compelling body of evidence which shows that women not only bear the brunt of poverty but that women's empowerment is a central precondition for its elimination. Poverty elimination can only be achieved by addressing the disproportionate burden of poverty, lack of access to education and health services, and lack of productive opportunities borne by women

- **Gender equality is integral to a rights-based approach to development.**
  Human rights, defined and upheld by the international community, are universal, and based on the equal worth and dignity of all women and men. Internationally agreed human rights include standards of health, education and the right to a secure livelihood, as well as civil, political and legal rights.

  Everywhere there are significant ways in which men's and women's responsibilities, opportunities and influence are unequal, although the nature and extent of inequality varies from society to society. Whilst there are instances where men are disadvantaged in comparison to women, generally women and girls have fewer opportunities, lower status and less power and influence than men and boys.

Millions of women around the world:
- have to work harder than men to secure their livelihoods
- have less control over income and assets
- have a smaller share of opportunities for human development
- are subject to violence and intimidation
- have a subordinate social position
- are poorly represented in policy-and decision making
- Gender inequality represents a huge loss of human potential, with costs for men as well as for women.

Δ What does gender equality mean?

Gender equality does not simply or necessarily mean equal numbers of men and women or boys and girls in all activities, nor does it necessarily mean treating men and women or boys and girls exactly the same. It signifies an aspiration to work towards a society in which neither women nor men suffer from poverty in its many forms, and in which women and men are able to live equally fulfilling lives. It means recognising that men and women often have different needs and priorities, face different constraints, have different aspirations and contribute to development in different ways.

Gender equality and women's empowerment are inextricably linked. Women will only win equality when they are able to act on their own behalf, with a strong voice to ensure their views are heard and taken into account. This means recognising the right of women to define the objectives of development for themselves.

The empowerment of women

The United Nations Development Fund for Women (UNIFEM) includes the following factors in its definition of women's empowerment:
- acquiring understanding of gender relations and the ways in which these relations can be changed
- developing a sense of self-worth, a belief in one's ability to secure desired changes and the right to control one's own life
- gaining the ability to generate choices and exercise bargaining power
- developing the ability to organise and influence the direction of social change to create a more just social and economic order, nationally and internationally.

An important message is that gender equality and the empowerment of women is achievable. Great progress has been made in the 20th century. Women enjoy greater freedom and more power than ever before. Progress has been greatest where there has been strong political will; where changes in laws, regulations and policies have been followed through with real action; where resources have been devoted to the explicit goal of reducing gender discrimination. Progress is not dependent on the income level of the society: some developing countries outperform much richer countries in the opportunities they afford women.

Another important message is that achieving gender equality is not a one-off goal. Progress can all too easily be eroded. Gender equality needs to be constantly promoted and actively sustained.

- Gender discrimination is any exclusion or restriction made on the basis of gender roles and relations that prevents a person from enjoying full human rights

Rural women suffer systematic discrimination in the access to resources needed for agricultural production and socio-economic development. Credit, extension, input and seed supply services usually address the needs of male household heads. Rural women are rarely consulted in development projects that may increase men's production and income, but add to their own workloads. When work burdens increase, girls are removed from school more often than boys, to help with farming and household tasks.

Δ Gender Equity

- Gender Equity is concerned with the promotion of personal, social, cultural, political, and economic equality for all. The term gender equity emerged out of a growing recognition in society of persuasive gender inequities.
- Discriminatory practices and continuing traditions of stereotypical conceptions have resulted in the systemic devaluation of attitudes, activities and abilities attributed to and associated with girls and women.
- The negative consequences of stereotypical conceptions and discriminatory practices adversely affect females as well as males.
- Gender equity means fairness and justice in the distribution of benefits and responsibilities between women and men. It often requires women-specific programmes and policies to end existing inequalities

Δ Gender Equity & Equality

- Gender Equity is the process of being fair to men and women. To ensure fairness, measures must often be put in place
- To compensate for the historical and social disadvantages that prevents women and men from operating on a level playing field.
- Equity is a means. Equality is the result.

- Equity: The distribution of rewards in society according to some criterion of merit, i.e., procedural justice and fairness. Outcomes reflect individual contributions.
- Equality: Giving the same reward to all, regardless of their contributions

Δ Gender sensitivity

- Gender sensitivity is the act of being aware of the ways people think about gender, so that individuals rely less on assumptions about traditional and outdated views on the roles of men and women.
- In language and the humanities, this is often expressed through people's language choice. People can choose more inclusive language that doesn't define gender, and many new words that are gender neutral have entered languages like English to substitute for more gender specific terms.
- It encompasses the ability to acknowledge and highlight existing gender differences, issues and inequalities and incorporates these into strategies and actions.

Δ Gender mainstreaming

Gender mainstreaming is the process of assessing the implications for women and men of any planned action, including legislation, policies or programmes, in all areas and at all levels. It is a strategy for making women as well as men concerns and experiences an integral dimension of the design, implementation, monitoring and evaluation of policies and programmes in all political, economic and societal spheres so that women and men benefit equally and inequality is not perpetuated. The ultimate goal is to achieve gender equality.

- It is the process of ensuring equal access and control over resources, development benefits and decision making at all stages of the development process for both men and women.
- Incorporation of a gender equality perspective in all development policies, strategies and interventions at all levels and at all stages by the actors normally involved therein.
- Considering both men's and women's wishes, needs, and experience in design, implementation, monitoring and evaluation of policies and efforts.

Δ Definition of Gender Mainstreaming

The concept of bringing gender issues into the mainstream of society was clearly established as a global strategy for promoting gender equality in the Platform for Action adopted at the United Nations Fourth World Conference on Women, held in Beijing (China) in 1995. It highlighted the necessity to ensure that gender equality is a primary goal in all area(s) of social and economic development.

In July 1997, the United Nations Economic and Social Council (ECOSOC) defined the concept of gender mainstreaming as follows:

"Mainstreaming a gender perspective is the process of assessing the implications for women and men of any planned action, including legislation, policies or programmes, in any area and at all levels. It is a strategy for making the concerns and experiences of women as well as of men an integral part of the design, implementation, monitoring and evaluation of policies and programmes in all political, economic and societal spheres, so that women and men benefit equally, and inequality is not perpetuated. The ultimate goal of mainstreaming is to achieve gender equality."

Mainstreaming includes gender-specific activities and affirmative action, whenever women or men are in a particularly disadvantageous position. Gender-specific interventions can target women exclusively, men and women together, or only men, to enable them to participate in and benefit equally from development efforts. These are necessary temporary measures designed to combat the direct and indirect consequences of past discrimination.

Transformation by Mainstreaming

Mainstreaming is not about adding a "woman's component" or even a "gender equality component" into an existing activity. It goes beyond increasing women's participation; it means bringing the experience, knowledge, and interests of women and men to bear on the development agenda.

It may entail identifying the need for changes in that agenda. It may require changes in goals, strategies, and actions so that both women and men can influence, participate in, and benefit from development processes. The goal of mainstreaming gender equality is thus the transformation of unequal social and institutional structures into equal and just structures for both men and women.

Why Gender Mainstreaming is important?

Gender mainstreaming ensures that policy-making and legislative work is of higher quality and has a greater relevance for society, because it makes policies respond more effectively to the needs of all citizens – women and men, girls and boys. Gender mainstreaming makes public interventions more effective and ensures that inequalities are not perpetuated.

Gender mainstreaming does not only aim to avoid the creation or reinforcement of inequalities, which can have adverse effects on both women and men. It also implies analysing the existing situation, with the purpose of identifying inequalities, and developing policies which aim to redress these inequalities and undo the mechanisms that caused them.

How does it work?

A political commitment for gender equality and a compatible legal framework are the basic conditions for the development of a successful gender mainstreaming strategy. In addition to concrete objectives and targets in the strategy, gender mainstreaming requires a clear action plan. Such plan should take into account the context, satisfy the necessary conditions, cover all the relevant dimensions, foresee the use of concrete methods and tools, set out the responsibilities and make sure that the necessary competences exist to achieve the anticipated results within a planned time frame.

Dimensions of gender mainstreaming

Gender mainstreaming requires both integrating a gender perspective to the content of the different policies, and addressing the issue of representation of women and men in the given policy area.

Both dimensions – gender representation and gender responsive content-need to be taken into consideration in all phases of the policy-making process.

Enabling conditions for gender mainstreaming

An effective implementation of gender mainstreaming requires preparation and organisation. People in decision-making positions can make a particular difference here, as they have more power to introduce changes.

Key elements to consider are:

- Preparation: set up a plan for the implementation of gender mainstreaming, define steps and milestones, assign tasks and responsibilities, formalise and communicate the plan.
- Resources: sufficient resources need to be made available; effective gender mainstreaming requires budget and time. Think about resources for awareness-raising and capacity-building initiatives. The use of special (external) expertise might also be considered.
- Stakeholder involvement: close liaison with all policy stakeholders is essential throughout the policy cycle to take on board the concerns, expectations, and views of the target groups. It is recommended to cement opportunities and structures for stakeholder involvement and consultations into the policy process.
- Monitoring and evaluation: set in place accountability mechanisms to ensure an adequate follow-up of implementation and progress. Foresee regular reporting and share results.
- Knowledge generation: building up knowledge on gender equality and good practices in gender mainstreaming contributes to making the approach more effective. You can contribute to the institutional learning by collecting data and information on indicators, reporting on progress and facilitating experience exchange.
- Gender expertise: this expertise should be internal, but the use of special external expertise might be considered as well.

Δ The four key steps of gender mainstreaming

Step 1: Sex disaggregated data and gender analytical information

Gender analytical research and sex disaggregated statistical data (about "beneficiary" groups and about management and implementation organisations) is essential to effective gender mainstreaming. Information systems should routinely be disaggregated by sex; gender analysis (an examination of women's as well as men's experiences, needs and priorities) should routinely be part of social and institutional appraisal and monitoring processes; and gender analytical studies should be commissioned to examine particular issues and address information gaps. This information is necessary to identify gender difference and inequality; to make the case for taking gender issues seriously; to design policies and plans that meet women's and men's needs; to monitor the differential impact of policy, project and budget commitments on women and men.

Step 2: Women as well as men influencing the development agenda

Women will only win equality when they are able to act on their own behalf, with a strong voice to ensure their views are heard and taken into account. This means promoting the involvement of women as well as men in decision-making at all levels, and ensuring that men and women committed to the promotion of gender equality are influencing decision-making. "Gender advocates" within government, civil society and donor organisations are most effective when they work in collaboration, identifying and developing strategic "entry points" for the promotion of gender equality.

Step 3: Context-specific action to promote gender equality

Gender mainstreaming is a strategy to promote gender equality and the empowerment of women. Action to promote greater equality of influence, opportunity and benefit should be devised on the basis of context-specific sex disaggregated data and gender analytical information and a clear understanding of women's and men's priorities. Actions need to be explicitly included in policy and project documents and frameworks, backed up with staff and budgets, and monitored and reviewed through appropriate indicators of change.

Step 4: Organisational capacity building and change

Gender mainstreaming, as an organisational strategy to promote gender equality, depends on the skills, knowledge and commitment of the staff involved in management and implementation. "Evaporation" of policy commitments to gender equality is widespread. Developing appropriate understanding, commitment and capacity, as well as addressing issues of gender inequality within development organisations themselves, is a long-term process of organisational change. Appropriate capacity-building activities need to be explicitly included in policy and project documents and frameworks, backed up with staff and budgets, and monitored and reviewed through appropriate indicators of change.

Δ Gender Relations

Definition of Gender relations: It is the relationship and power distribution between women and men in a given socio-cultural context. It indicates ways in which a culture or society defines rights, responsibilities and identities of men and women in relation to one another.

- Women and Men are the two important factors in the society. Gender division of labour establishes gender relations in the families and in the societies therefore these relations can either be based on relationship of cooperation or relationship of conflict.
- Society needs to be gender friendly and gender sensitive. This reduces each other's vulnerability and would lead to understanding of gender needs in a better way.
- Gender does not only describe the different roles and relationships of women and men but also the power structures inherent in their relationships. This is due to 'patriarchy'.

Δ Sex-disaggregated statistics

The collection and separation of data and statistical information by sex to enable comparative analysis; sometimes referred to as gender-disaggregated statistics.

Δ Special interventions

Special interventions are efforts aimed at creating fundamental structural changes in institutions, policies, legislation, and allocation of resources to promote gender equality between men and women, based on the specific needs in the individual country, policy area or organization.

Δ Women's empowerment

- The empowerment of women concerns women gaining power and control over their own lives
- It constitutes an important part of the efforts to bring about equal opportunities for men and women
- and involves awareness raising, building self-confidence, expansion of choices, increased access to and control over resources and
- Actions to transform the structures and institutions which reinforce and perpetuate gender discrimination and inequality.
- Gender Division of Labour-is the result of how each society divides work among men and among women according to what is considered suitable or appropriate to each gender.

Δ Women's rights

The rights of women and the girl child are an inalienable, integral, and indivisible part of universal human rights.

Δ De facto women-headed household

These are women-headed households in which the male partner is 'temporarily' absent. This is mostly due to male labour out migration and the women is not the legal household head.

Δ De jure female-headed household

These are women-headed household in which the male partner is 'permanently' absent due to separation or death, and the woman is legally single, divorced or widowed.

Δ Gender blind

Gender blind is a person who does not recognize that gender is an essential determinant of life choices available to people in society. A gender blind approach assumes that gender is not an influencing factor in projects, programmes or policy matter.

Ignoring/failing to address the gender dimension (as opposed to gender sensitive or gender neutral)

Δ Gender neutral

Gender neutral policies are not specifically aimed at either men or women and are assumed to affect both sexes equally. However, they may actually be gender blind.

Δ Gender bias

Perception that both sex are not equal and do not have similar rights to resources.

Δ Gender discrimination

It refers to un-favorable treatment of individuals on the basis of their gender. It is the pre-judicial treatment of an individual based on a gender stereotype (often referred to as sexism or sexual discrimination).

Δ Gender Issues

It refers to specific consequences of the inequality between women and men.

Δ Gender Norms

- Gender norms are the accepted attributes and characteristics of male and female gendered identity at a particular point in time for a specific society or community.
- They are the standards and expectations to which gender identity generally conforms, within a range that defines a particular society, culture and community at that point in time.
- Gender norms are ideas about how men and women should be and act. Internalized early in life, gender norms can establish a life cycle of gender socialization and stereotyping.

Δ Gender needs and assessment

- Women and men have different roles and responsibilities therefore, they have different interests and needs.
- Some needs, which you consider essential for men, may not be so for women. Women have particular needs that differ from those of men, are only because of their triple role, but also because of their subordinate position in terms of men. Moreover, the priorities of needs to women and men also differ.
- However, women men also have common needs. They have similar emotions and feelings of happiness and pain and suffer from similar diseases. Needs of food, house, cloth and water, therefore, are common to them.

Δ Types of Gender Needs

According to Molyneux, there are two types of Gender Needs, these are:
1) Practical Gender Needs
2) Strategic Gender Needs

1) Practical Gender Needs (PGN):

Refers to the needs which are derived from women and men's position within the gender division of labor.

Practical Gender Needs of food, house, cloth, water and health are same for both women and men

2) Strategic Gender Needs (SGN):

- Those needs that arise from changes in the gender division of labor e.g. when women take on work not traditionally seen as women's work and men take more responsibility for child care and domestic work.
- Giving specific protection to women under legal rights, an end to violence both domestic, at work-place and in nations, equal wages to women and men and having control over their own bodies.

- Gender planning refers to the process of planning developmental programmes and projects that are gender sensitive and which take into account the impact of differing gender roles and gender needs of women and men in the target community or sector. It focuses on gender with an objective to achieve gender equity, equality and empowerment through practical and strategic gender needs.

- Gender-responsive objectives are programme and project objectives that are non-discriminatory, equally benefit women and men and aim at correcting gender imbalances.

- Gender Parity: It is a numerical concept. Gender parity concerns relative equality in terms of numbers and proportions of men and women, girls and boys. Gender parity addresses the ratio of female to male values (or males to females, in certain cases) of a given indicator.

- Literacy Gender Parity Index (GPI) is the ratio of the female to male adult literacy rates which measures progress towards gender equity in literacy and the level of learning opportunities available for women in relation to those available to men. It serves also as a significant indicator of the empowerment of women in society.

- Gender Perspective: Using a 'gender perspective' entails approaching or examining an issue, paying particular attention to the potentially different ways in which men and women are or might be affected.

This is also called using or looking through a 'gender lens.'

Δ Gender awareness:

The knowledge and understanding of the differences in roles and relations between women and men, especially in the workplace: It is important to continuously raise gender awareness and understanding among government officials and the public in general.

Δ Glass Ceiling:

It refers to impediments that prevent women from rising to top positions in an organization, including the political, public, private, judicial, social, and economic domains.

The term "glass" is used as these impediments are apparently invisible and are usually linked to the maintenance of the status quo in organizations as opposed to transparent and equal career advancement opportunities for women and men within organizations

Δ Gender balance:

Gender balance is commonly used in reference to human resources and equal participation of women and men in all areas of work, projects and programmes.

In a scenario of gender equality, women and men are expected to participate proportionally to their share of the population. In many areas, however, women participate less than what would be expected based on the sex distribution in the population (underrepresentation of women), while men participate more than expected (overrepresentation of men).

Δ Women's Rights:

- Rights that promote a position of legal and social equality of women with men. These are the rights and entitlements claimed for women and girls of many societies worldwide, and formed the basis to the women's rights movement in the nineteenth century and feminist movement during the 20th century.

- In some countries, these rights are institutionalized or supported by law, local custom, and behaviour, whereas in others they may be ignored or suppressed.

- Issues commonly associated with notions of women's rights include, though are not limited to, the right: to bodily integrity and autonomy; to be free from sexual violence; to vote; to hold public office; to enter into legal contracts; to have equal rights in family law; to work; to fair wages or equal pay; to have reproductive rights; to own property; to education.

Δ Sex:

"Sex" refers to the biological differences between women and men. They are generally permanent and universal.

Sex, is a biological term referring to people, animals, etc., being either female or male depending on their sex organs or genes. Sex also refers to the differences between individuals that make them male or female. These differences are biologically determined.

- Gender Mainstreaming: Gender mainstreaming is the process of ensuring that women and men have equal access and control over resources, development benefits and decision-making, at all stages of the development process and UNDP projects, programmes and policy.

- Gender-sensitivity: Gender sensitivity encompasses the ability to acknowledge and highlight existing gender differences, issues and inequalities and incorporates these into strategies and actions.

- Gender Discrimination: Prejudicial treatment of an individual based on a gender stereotype (often referred to as sexism or sexual discrimination).

- Gender equity: Gender equity entails the provision of fairness and justice in the distribution of benefits and responsibilities between women and men. The concept recognises that women and men have different needs and power and that these differences should be identified and addressed in a manner that rectifies the imbalances between the sexes.

(OR)

Condition in which women and men participate as equals, have equal access to resources, and equal opportunities to exercise control.

- Gender division of labour: The roles, responsibilities, and activities assigned to women and men based on gender.
- Gender equality: Gender equality is the result of absence of discrimination on the basis of a person's sex in opportunities and the allocation of resources or benefits or in access to services.
- Gender Issues: Specific consequences of the inequality between women and men.
- Gender relations: Ways in which culture or society defines rights, responsibilities, and identities men and women in relation to one another. Gender disaggregated data:

For a gender analysis, all data should be separated by sex / gender in order to allow differential impacts on men to be identified, seen and measured.

- Gender planning: Gender planning refers to the process of planning developmental programmes and projects that are gender sensitive and which take into account the impact of differing gender roles and gender needs of women and men in the target community or sector.
- Gender disaggregated data: For a gender analysis, all data should be separated by sex / gender in order to allow differential impacts on men to be identified, seen and measured.

Sex disaggregated data is quantitative statistical information on differences and inequalities between women and men.

Sex disaggregated data might reveal, for example, quantitative differences between women and men in morbidity and mortality; differences between girls and boys in school attendance, retention and achievement; differences between men and women in access to and repayment of credit; or differences between men and women in voter registration, participation in elections and election to office.

Δ Gender budgeting:

A gender budget is not a separate budget for women. Instead, the gender budgets are an attempt to assess government priorities as they are reflected through the budget and examine how they impact women and men.

Gender budgets look at what the impact of spending is on men and women and whether or not budgets respond to the needs of both women and men adequately.

"Women's budgets", "gender budgets", "gender-sensitive budgets", and "gender responsive budgets" are all terms that are used to describe initiatives that have used gender as lens from which to analyse budgets at national, regional, and civic levels.

"Gender budget initiatives analyse how governments raise and spend public money, with the aim of securing gender equality in decision-making about public resource allocation; and gender equality in the distribution of the impact of government budgets, both in their benefits and in their burdens. The impact of government budgets on the most disadvantaged groups of women is a focus of special attention."

The term gender budgeting has been defined differently in various documents on the subject. A comprehensive definition as:"Gender budgeting is a dissection of the government to establish its gender differential impacts and to translate gender commitments into budgetary commitments" Thus gender budgeting looks at the government budget from a gender perspective to assess how it addresses the needs of women in the areas like health, education, employment, etc.., gender budgeting does not seek to create a separate budget but seeks affirmative action to address specific needs of women. Gender responsive budgeting initiatives provide a way of assessing the impact of government revenue and expenditure on women.

Gender Responsive Budget

A Gender-Responsive Budget is a budget that acknowledges the gender patterns in society and allocates the money to implement policies and programs that will change these patterns in a way that moves towards a more gender equal society. Gender budget initiatives are exercises that aim to move the country in the direction of a gender-responsive budget.

Need of a Gender Budget

Gender Budget Initiatives are attempts to disaggregate the government's mainstream budget according to its impacts on women and men. It refers to the process of conceiving, planning, approving, executing, monitoring, analysing and auditing budgets in a gender-sensitive way. The gender budgeting exercise would potentially assist and lead to the following empowering measures:

- Addressing gap between policy commitment and allocation for women by emphasizing on adequate resource allocation.
- Putting pressure and focus on gender sensitive programme formulation and implementation.
- Mainstreaming gender concerns in public expenditure and policy.
- By being a tool for effective policy implementation where one can check if the allocations are in line with slated gender sensitive policy commitments and are having the desired impact.

Gender budget is helpful in

- Improving women's economic equality.
- Improving effectiveness, efficiency, accountability, and transparency of government budgets.
- Revealing discrepancies between what a governments says it is doing and the actual impact of government policies.
- Offering a practical way for the governments to implement their obligations under international human rights agreements such as the Convention on the Elimination of All Forms of Discrimination against Women (CEDAW).

Why Gender budgeting?

Budgeting is universally accepted as a powerful tool in achieving development objectives and act as an indicator of commitment to the stated policy of the government. National budgets reflect how governments mobilize and allocate public resources, and how they aim to meet the social and economic needs of their people. The budgetary policy of the government has a major role to play in achieving objectives of gender equality and growth through content and direction of Fiscal and Monetary policies, measures for resource mobilization, affirmative action for under privileged sections etc. women stand apart as one segment of the population that warrants special attention due to their vulnerability and lack of access to state resources. Thus gender responsive budgets policies can contribute to achieving the objectives of gender equality, human development and economic efficiency. The purpose of gender budgeting exercise is to assess quantum and adequacy of allocation of resources for women and establish the extent to which gender commitments are translated in to the budgetary commitments. This exercise facilitates increase in accountability, transparency and participation of the community. Gender mainstreaming requires gender responsive policy. When gender equality considerations are incorporated into policy making, the concerns and needs of both women and men become integral part of the design, implementation, monitoring and evaluation of policies and programmes in all sections of society.

The Scope of Gender Budgeting

Gender Budgeting expands our concept of the economy to include things that are not usually valued in money. In particular, Gender Budgeting recognizes the unpaid care economy-the work that mainly women do in bearing, rearing and caring for their families and the people in our society. Gender Budgeting recognises that unless this unpaid care work is done, the economy will not function effectively and people's well-being will be very negatively affected. Government therefore needs to find ways of supporting those who do this unpaid care work, lessening their burden, and ensuring that the work is done well. Gender Budgeting should, however, not be confined to the 'social' or 'soft' areas such as education, health and welfare. Gender Budgeting is a tool for gender mainstreaming in the developmental process as a whole. As such, it needs also to be applied in areas such as agriculture, power, defence, commerce, and information technology where the gender implications may not be immediately apparent. Later chapters give examples of how this can be done.

Δ Major characteristics

- Refers to the process of conceiving, planning, approving, executing, monitoring, analyzing and auditing budgets in a gender-sensitive way.
- An exercise to translate stated gender commitments of the Government into budgetary commitments.
- Strategy for ensuring Gender Sensitive Resource Allocation and a tool for engendering macroeconomic policy.
- Covers assessment of gender differential impact of Government Budgets and policies (Revenue and Expenditure).
- Enables Tracking and Allocating resources for women empowerment.
- Opportunity to determine real value of resources allocated to women.

Gender Budget Cells

The Department of Expenditure, Ministry of Finance issued a Charter for the Gender Budget Cells, on 8 March 2007, clearly articulating the composition and the functions of the gender budget cells. The MWCD has been actively pursuing other Ministries with regard to setting up of gender budget cells. These gender budget cells serve as focal points for coordinating Gender Budgeting initiatives both intra-and inter-ministerially. The roles envisaged for these cells include:

Role of Gender budgeting cells

- Act as a nodal agency for all gender responsive budgeting initiatives.
- Pilot action on gender sensitive review of public expenditures and policies.
- Guide and undertake collection of gender disaggregated data-for target group of beneficiaries covered under expenditure, revenue rising/ policy/ legislation.
- Guide gender budgeting initiatives within department as well as in field units responsible for implementing government programmes.
- Conduct gender based impact analysis, beneficiary needs assessment and beneficiary incidence analysis to
  1. Establish effectiveness of public expenditures.
  2. Identify scope for reprioritization of public expenditure.
  3. Improve implementation etc.
  4. Collect and promote best practices on participative budgeting for and implementation of schemes.

Δ Guidelines for Gender Sensitive Budgeting

- Preparation of Gender based profile of public expenditure
- Beneficiary Needs Assessment
- Impact Analysis of public expenditure and policies-policy & programme design change in quantum of allocation implementation guidelines
- Beneficiary Incidence Analysis
- Participative Budgeting

Δ Tools of Gender Budgeting

Gender budget tools

- Gender-aware policy appraisal: this is the most common approach. It begins with the assumption that budgets reflect policy. Analysis involves scrutinising the explicit and implicit gender implications of national and sectoral policies, examining the ways in which priorities and choices are likely to reduce or increase gender inequality.
- Gender-disaggregated beneficiary assessments: this is a more participatory approach to policy analysis i.e. asking actual or potential beneficiaries the extent to which government policies/programmes match their own priorities.
- Gender-disaggregated public expenditure incidence analysis: this compares public expenditure for a given programme with data from household surveys to reveal the distribution of expenditure between women/men, boys/girls.
- Gender-disaggregated tax incidence analysis: this examines direct and indirect taxes and user fees to calculate how much tax is paid by different individuals and households.
- Gender-disaggregated analysis of the impact of the budget on time use: this examines the relationship between national budget and the way time is used in households. In particular, it draws attention to the ways in which the time spent by women in unpaid work is accounted for in policy analysis.
- Gender-aware medium term economic policy framework: these are attempts to incorporate gender into the economic models on which medium term economic frameworks are based.
- Gender-aware budget statement: this is an exercise in government accountability which may use any of the above tools. It requires a high degree of commitment and co-ordination throughout the public sector as ministries and departments undertake and publicise an assessment of the gender impact of their line budgets.

Five Steps Framework for Gender Budgeting

Step 1	An analysis of the situation for women and men and girls and boys (and the different sub groups) in a given sector
Step 2	An assessment of the extent to which the sector's policy addresses the gender issues and gaps described in the first step. This step should• include an assessment of the relevant legislation, policies, programmes and schemes. It includes an analysis of both the written policy as well as the implicit policy reflected in government activities. It should examine the extent to which the above meet the socio-economic and other rights of women.
Step 3	An assessment of the adequacy of budget allocations to implement. The gender sensitive policies and programmes identified in step 2 above.
Step 4	Monitoring whether the money was spent as planned, what was delivered and to whom. This involves checking both financial performance and the physical deliverables (disaggregated by sex)
Step 5	An assessment of the impact of the policy / programme / scheme and the extent to which the Situation described in step 1 has been changed, in the direction of greater gender equality

Source: (UNIFEM-UNFPA Gender Responsive Budgeting and Women's Reproductive Rights: Resource pack)

Case Study: Gender Budgeting In Agriculture

This case study is based on research conducted for the Gender Resource Centre of the Ministry of Agriculture, GOI by Neeraj Suneja.

Rural women are major producers of food in terms of value, volume and hours of work. Nevertheless, women's control over resources and processesn remains extremely limited. Women may function as head of the household for the major part of the year. Nevertheless, the landlords and officials continue to recognise the husband as the cultivator in the official lists and statistics. Women then have difficulty in accessing credit and inputs from mainstream institutions and government schemes and also in becoming members of farmers associations and beneficiary organisations. Agricultural research has also focused on increasing the production of high value major cereal and cash crops rather than the traditional varieties of cereals and subsistence crops which are farmed by women and which provide the major source of food to their families. Upgrading of technology has focused on implements and tools designed with male users in mind.

The National Agriculture Policy of 2000 gave high priority to 'recognition and mainstreaming of women's role in agriculture'. At state level, states are encouraged to allocate 30% of allocations for women farmers and women extension functionaries under the extension interventions, focusing on formation of Women

SHGs; capacity building interventions; linking women to micro credit; and improving their access to information through IT and other extension activities. At the central level, a National Gender Resource Centre in Agriculture (NGRCA) has been established to assist the centre and the states with advisory services.

The Ministry of Agriculture has started a number of programmes and schemes which target women. These include:

Horticulture: The State Horticulture Missions have been directed to earmark at least 30% of their budgets for women beneficiaries in all ongoing programmes under the National Horticulture Mission and Technology Mission for Horticulture in North Eastern States, Sikkim, Jammu & Kashmir, Himachal Pradesh and Uttaranchal.

Agriculture Extension: In the scheme "Support to States Extension Programme for Extension Reforms", 30% of resources are meant to be allocated for women farmers and extension functionaries.

Watershed Development Programmes: The Watershed Development programmes provide for the involvement of women farmers in the constitution of Watershed Associations and other institutional arrangements and formation of women SHGs and User Groups (UGs).

Crops: The scheme "Technology Mission on Cotton" encourages states/ implementing agencies to give preference to women farmers in components like distribution of agriculture inputs, trainings and demonstrations so that at least 20% of the total allocation reaches them. Under another Mission, a subsidy is provided for the distribution of sprinkler sets to women farmers and other disadvantaged groups.

Technology Mission on Oilseeds & Pulses: The "Integrated Scheme of Pulses, Oilseeds, Palm Oil and Maize" provides subsidy/assistance to women farmers for sprinkler sets and pipes for carrying water from source to the field.

Integrated Nutrient Management: 25% of seats are reserved for women in the training courses for farmers on organic farming.

Some States have also initiated schemes targeting women. For example, a scheme launched in the Tenth Plan by the Extension Division namely "Support to States for Extension Reforms" provides for representation of women in all bodies at district level, including the governing board, farmer advisory committees, farm women interest groups and commodity-based organisations.

Different players / stakeholders in Gender budgeting initiatives

There are a range of different actors who can be involved in Gender Budgeting. They have different roles and carry out different activities. Some of them are:

- The Ministry of Women and Child Development(nodal ministry at the central level, in India)
- The Ministry of Finance (at the Centre and in the States)
- The Planning Department or Planning Commission (at the Centre and in the States)

- Sectoral ministries-each and every department and ministry can do Gender Budgeting – albeit some have more possibilities than others (at the Centre and in the States)
- Researchers and economists
- Statisticians
- Civil society organizations especially women's groups
- Parliamentarians, MLAs and other representatives of the people at state/ district and sub-district levels.
- Media
- Development partners/donors
- The women and men for whom the specific policy, programme or budget is intended

Difference between Gender Budgeting and Gender Auditing

Gender auditing is part of the Gender Budgeting process. Gender auditing is the process that is conducted after the budget has been implemented. It is the process of reviewing financial outlays-looking at trends over time, percentage shares etc; analysing and assessing systems actually put in place, processes adopted, outcomes and impacts of budgetary outlays vis-à-vis what was planned – all this through a gender lens.

Δ Assessing Gender Empowerment

- Basic socio-economic infrastructure-Health, Education, Water & sanitation
- Economic Empowerment-Economic identity, Employment, Assets, Credit, Skills, Markets, Risk coverage
- Social and political empowerment-Political participation, Gender Equality in inheritance, marital laws, security.
- Action areas: Spatial Maps & Yardsticks for Availability of Resources and Opportunities. Synergy in Resource allocation across levels of governance to ensure universal coverage. Redesign programmes from gender perspective-build in women's participation-break gender barriers in access to public expenditure. Planning for access to Sustained Employment in all habitations and Social Security. Capacity Building for Women-budgeting, political processes, leadership, collective power. Training in high end skills and entrepreneurship for more productivity and remuneration. Collection of gender dis-aggregated data.
- Gender Sensitive Finances: Translate gender based spatial requirements in to resource allocations. Re-prioritize resource allocations to address-regional imbalances, infrastructure gaps. Gender perspective on revenue raising policies/subsidies. Gender perspective in monetary and fiscal policies.
- Gender Sensitive Administration: Better monitoring of programmes-based on spatial progress, better implementation of laws, Better Fiscal

Management, Social re-engineering-Education at all levels, Training & Gender sensitization of administrative cadres , Bridging gap between Research and Administration-Evaluations, Surveys, Research.

- Gender issues: specific consequences of the inequality between women and men
  1. Traditionally women are expected to perform work which involves manual, respective tasks.
  2. Men posses and control land
  3. Women's work is underpaid and undervalued
  4. Women's work is unaccounted
  5. Women undertake multiple workload at home, farm and community.
  6. Women lack access to skills, trainings and information
  7. Women get displaced from traditional employment opportunities.
  8. When technology is introduced and men take over the task
  9. Women lacks access to firm inputs, credit and financial incentives
  10. Women lacks access to market and control over income
  11. Women face food insecurity
  12. Women's priorities, needs, problems are unattended and neglected
  13. Farmer is regarded as male
  14. Women are discriminated in design and testing of technologies.
  15. Feminization of Agriculture
  16. Over-burden of Work
  17. Impact of Technology
  18. Facilities and Support Services
  19. Development Bias
  20. Constraints to Women's Access to Resources
  21. Access to Land
  22. Access to Credit
  23. Access to Markets
  24. Research and Technology Development
  25. Access to Extension and Training
  26. Training and Capacity Building of Farm Women

Δ Challenges for development of Gender
  1. Gender stereotypes
  2. Gender bias
  3. Gender discrimination
  4. Gender division of labour

Δ **To overcome from this challenges we need to take up the following methods**

1. Gender sensitization
2. Gender mainstreaming
3. Gender budgeting
4. Gender equity
5. Gender equality

Δ **Advantages from the these methods:**

1) Looking at gender differences by identifying tasks, activities and rewards associated with the gender division of labour
2) Bridging agricultural technological gap between women and men farmers
3) Bridging agricultural information gap
4) Focusing on agricultural interventions from gender perceptive
5) Examining budget from a gender perceptive
6) Looking into programmes / project activities from a gender lens
7) Developing action plans from a gender perceptive
8) Mainstreaming gender into developmental activities including training, transfer of technology mechanism and other extension services.

   • Gender Stereotypes: refer to oversimplified and standardised gender concepts that are commonly held by members of a group. Until the 1960s these were perceived as being biological in origin. With the birth of the feminist and human rights movements, gender stereotypes were perceived as being socially constructed rather than biologically determined. Some examples of stereotypes are that males are competitive, strong, aggressive and independent whereas females are sensible, sweet, submissive and dependent.

   • Division of labour (by gender)-The division of paid and unpaid work between women and men in private and public sphere.

   • Empowerment: The process of gaining access and developing one's capacities with a view to participating actively in shaping one's own life and that of one's community in economic, social and political terms.

   • Equal opportunities for women and men: The absence of barriers to economic, political and social participation on the grounds of sex.

   • Equal pay for work of equal value: Equal pay for work to which equal value is attributed without discrimination on grounds of sex or marital status with regards to all aspects of pay and conditions of remuneration.

   • Family responsibilities: Family responsibilities cover the care of and support for dependent children and other members of the immediate family who need help. National policies should aim at creating effective equality opportunity and treatment for female and male workers, and for workers without family responsibilities. Such policies should be free from restrictions based on family responsibilities when preparing for and entering, participating in or advancing in economic activity.

- Feminisation of poverty: The increasing incidence and prevalence of poverty among women compared to men.

- Gender gap: The gap in any area between women and men in terms of their levels of participation, access, rights, remuneration or benefits.

- Gender contract: A set of implicit and explicit rules governing gender relation which allocate different work and value, responsibilities and obligations to men and women and is maintained on three levels: cultural superstructure – the norms and values of society; institutions – family welfare, education training, equal pay for work of equal value, rights to land and other capital assets, prevention of sexual harassment at work and domestic violence, and freedom of choice over childbearing. Addressing them entails a slow transformation of the traditional customs and conventions of a society.

- Gender neutral: Having no differential positive or negative impact for gender relations or equality between women and men.

- Human rights of women: The rights of women and the girl child as inalienable, integral and indivisible part of universal human rights.

- Gender impact assessment: Examining policy proposals to see whether they will affect women and men differently, with a view to adapting these proposals to make sure that discriminatory effects are neutralised and that gender equality is promoted.

- Occupational (job) segregation: The concentration of women and men in different types and levels of activity and employment, with women being confined to a narrower range of occupations (horizontal segregation) than men, and to the lower grades of work (vertical segregation).

- National women's (gender) machinery: A national machinery for the advancement of women is the central policy-co-ordinating unit inside government. Its main task is to support the government wide mainstreaming of a gender equality perspective in all policy areas.

- Participation rates: The rate of participation by defined groups – example: women, men, lone parents, etc. – as a percentage of overall participation, usually in employment.

- Men and masculinities: Addressing men and boys refers to better understanding the male side of the gender equation. It involves questioning the masculine values and norms that society places on men's behaviour, identifying and addressing issues confronting men and boys in the world of work, and promoting the positive roles that men and boys can play in attaining gender equality.

- Informal economy/work: Unpaid economic activities done for the direct benefit of the household or of relations' and friends' households on a reciprocal basis, including everyday domestic work and a great variety of self provisioning activities and/or professional activity, whether as a sole or secondary occupation, exercised gainfully and not occasionally, on the limits of, or outside, statutory, regulatory or contractual obligations, but excluding informal activities which are also part of the criminal economy.

- Reproductive rights: The right of any individual or couple to decide freely and responsibly the number, spacing and timing of their children and to have the information and means to do so, and the right to attain the highest standard of sexual and reproductive health.

- Sex: The biological characteristics which distinguish human beings as female or male.

- Sex discrimination – direct: Where a person is treated less favourably because of his or her sex.

- Sex discrimination – indirect: Where a law, regulation, policy or practice, apparently neutral, has a disproportionate adverse impact on the members of one sex, unless the difference of treatment can be justified by objective factor.

- Sexual harassment: Unwanted conduct of a sexual nature or other conduct based on sex affecting the dignity of women and men at work including conduct of superiors and colleagues.

- Stereotypes: A fixed idea that people have about what someone or something is like, especially an idea that is wrong.

- Human development: Human development is about people, about expanding their choices to lead lives they value. Economic growth, increased international trade and investment, technological advance – all are very important. But they are means, not ends. Whether they contribute to human development in the 21st century will depend on whether they expand people's choices, whether they help create an environment for people to develop their full potential and lead productive, creative lives.

- Women's triple role: Women's triple role refers to the reproductive, productive and community managing role. The way these forms are valued affects the way women and men set priorities in planning programmes or projects. The taking or not taking into consideration of these forms can make or break women's chances of taking advantage of development opportunities.

- Equality of opportunity: this means that women should have equal rights and entitlements to human, social, economic and cultural development, and an equal voice in civic and political life.

- Equity of outcomes: this means that the exercise of these rights and entitlements leads to outcomes which are fair and just.

- Discrimination (direct and indirect): Discrimination occurs in various forms in everyday life. As defined by the ILO (2003). Any distinction, exclusion or preference made on the basis of race, colour, sex, religion, political opinion, national extraction or social origin which has the effect of nullifying or impairing equality of opportunity and treatment in employment or occupation is discriminatory. Alongside racial discrimination, gender discrimination can be seen as one major form of discrimination. Discrimination can be distinguished into two forms: direct and indirect. The first form arises if, without being less qualified, certain groups of society are explicitly excluded or disadvantaged by the legal framework due to characteristics such as gender. Indirect discrimination occurs if intrinsically neutral rules or laws negatively affect certain groups, e.g. female workers. Discrimination of part-time workers against full time employees is still present in nearly every country. As a major proportion of part-time workers are female, this disadvantages women as well.

- Discrimination (Gender): The Convention on the Elimination of all Forms of Discrimination Against Women (CEDAW), approved by the United Nations in 1979, states that "Discrimination against women shall mean distinction, exclusion, or restriction made on the basis of sex which has the purpose of impairing or nullifying the recognition, enjoyment or exercise by women, irrespective of their marital status, on a basis of equality of men and women, of human rights and fundamental freedoms in the political, economic, social, cultural, civil or any other field". It refers to any distinction, exclusion or restriction made on the basis of socially constructed 6 gender roles and norms, which prevents a person from enjoying full human rights.

- Discrimination (Systemic): Systemic discrimination is caused by policies and practices that are built into the ways that institutions operate, and that have the effect of excluding women and minorities. For example, in societies where the belief is strong that whatever happens within the household is the concern of household member only, the police force and judiciary, organisations within the institution of the state are likely routinely to avoid addressing questions of domestic violence, leading to systemic discrimination against all the women who experience violence within the home.

- Displaced Women: Displaced persons are those who have fled or been driven from their communities to other localities within their country of nationality According to the UNHCR, more than 75% of displaced persons are women and their children, they are subjected to physical and sexual violence as much during their flight as when they arrive in the country of asylum, be it from members of the armed forces, immigration agents, bandits, pirates, local populations, individuals belonging to rival ethnic groups or other refugees.

- Domestic Work: Work done primarily to maintain households. Domestic includes the provision of food and other necessities, cleaning, caring for children and the sick and elderly, etc. Domestic work is mostly performed by women and is therefore poorly valued in social and economic terms.

- Domestic Worker: In certain countries, in order to overcome the problem of a lack of childminding and/or care facilities, another type of female labour is used, namely domestic workers, mainly women, often immigrants sometimes undocumented and often under-paid. The demand for domestic workers is growing in the EU as a result of changes in the economy and society. In many situations, it has become necessary for households to employ women who are migrant workers so as to allow the parents who employ them to be active in the work place and in society.

- Equality of Outcome: Is sometimes also referred to as "substantive equality", and refers to the insight that equality of opportunity may not be enough to redress the historical oppression and disadvantage of women. Because of their different positions in society, women and men may not be able to take advantage of equal opportunities to the same extent. In some cases equal opportunities can actually have a negative impact on women's well-being, if women exert time and energy to take advantage of them with no result. In order to ensure that development interventions result in equality of outcome for women and men, it is necessary to design them on the basis of gender analysis. "Equal" treatment therefore does not mean "the same" treatment.

- Equity (and Sustainable Development): Equity derives from a concept of social justice. It represents a belief that there are some things that people should have, that there are basic needs that should be fulfilled, that burdens and rewards should not be spread too divergently across the community, and that policies should be directed with impartiality, fairness and justice towards these ends. Equity means that there should be a minimum level of income and environmental quality below which nobody falls. The central ethical principle behind sustainable development is equity and particularly intergenerational equity defined as a development that meets the needs of the present without compromising the ability of future generations to meet their own needs.

- Epistemology: An epistemology is a theory of knowledge. Feminists and gender researchers have argued that traditional epistemologies exclude the possibility that women can be "knowers" or agents of knowledge; they claim that the voice of science is a masculine one and that history is written exclusively from the point of view of men (of the dominant class and race).They have proposed alternative epistemologies that legitimate women's knowledge.

- Beijing Declaration and Platform for Action: The Beijing Declaration and Platform for Action was adopted by the Fourth World Conference on Women: Action for Equality, Development and Peace, Beijing, 15 September 1995. The document was agreed upon world governments at the Conference and is a comprehensive outline of strategic steps to be taken in order to concretise and enhance the goals of CEDAW (See CEDAW). Although it is not, of its nature, a legally binding document, consisting of policy commitments rather than legal obligations, it is, nonetheless, a significant statement of principle, and has great symbolic value.

- Androgyny: A term that combines the Greek words for man and woman, is a state of ambiguous gender in which identifying sexual characteristics are uncertain or mixed. It differs from hermaphroditism, or intersexuality, a condition in which dual sexual characteristics are unambiguously present. To say that someone is andryogynous is to say that he or she combines stereotypically male and female attributes.

- Bisexual: A person for whom their sexual attraction is more or less equally directed to a person of either sex.

- Capabilities Approach: Developed by economist and development expert Amartya Sen, the capabilities approach views the end goal of development as the expansion of the freedom of people to choose the kind of life they wish to live. Capabilities are "substantive human freedoms" – rather than focusing on income and wealth, they ask what choices people have, and what individuals are actually able to do and be. According the United Nations Development Programme (UNDP), which has outlined the approach in its Human Development Reports (see HDR 2000), there are three essential capabilities: for people to lead long and healthy lives, to be knowledgeable, and to have a decent standard of living. Martha Nussbaum has written about the relevancy of the capabilities approach to women, contending that "women's issues have been at the heart of the approach from the start, both because of their urgency and because the dire situation of women around the world helps us to see more clearly the inadequacy of various other approaches to development." Naila Kabeer has also recently included the approach in her examination of gender mainstreaming in poverty eradication and the Millennium Development Goals. Although Kabeer believes that improvements can be made, she 3 notes that the approach has in many respects been more successful in revealing the gender dimensions of poverty than other commonly used approaches.

- Care Work: Care work encompasses care provided to dependent children, the elderly, the sick and the disabled in care institutions or in the home of the person requiring care. Care policies and the provision of care services are intrinsically related to the achievement of equality between women and men. The lack of affordable, accessible and high quality care services and the fact that care work is not equally shared between women and men have a direct negative impact on women's ability to participate in all aspects of social, economic, cultural and political life.

- Care (Informal): Unpaid care for dependent children, the elderly, ill or disabled persons carried out by family members or others. The responsibility of informal care work is taken up by women with major impact on their health and well being. Informal care is largely invisible and the economic and social contributions of women carers unacknowledged. Over 75% of informal carers worldwide are women.

- CEDAW (The Convention on the Elimination of All Forms of Discrimination against Women): An international convention adopted in 1979 by the UN

General Assembly, is often described as an international bill of rights for women. Consisting of a preamble and 30 articles, it defines what constitutes discrimination against women and sets up an agenda for national action to end such discrimination. By accepting the Convention, States commit themselves to undertake a series of measures to end discrimination against women in all forms, including: • to incorporate the principle of equality of men and women in their legal system, abolish all discriminatory laws and adopt appropriate ones prohibiting discrimination against women; • to establish tribunals and other public institutions to ensure the effective protection of women against discrimination; and • to ensure elimination of all acts of discrimination against women by persons, organizations or enterprises. Countries that have ratified or acceded to the Convention are legally bound to put its provisions into practice. They are also committed to submit national reports, at least every four years, on measures they have taken to comply with their treaty obligations. Optional Protocol to CEDAW 4 was adopted in 1999 by the General Assembly. States which ratify the Optional Protocol recognize the competence of the Committee on the Elimination of Discrimination against Women to consider petitions from individual women or groups of women who have exhausted all national remedies. The Optional Protocol also entitles the Committee to conduct inquiries into grave or systematic violations of the Convention.

- Customary Law: In many countries, a system of civil law runs parallel to indigenous and religious systems of customary law. Customary law often applies in matters concerned with family law, and thus as a great deal of impact on women's everyday lives, as it deals with issues such as marriage, divorce, inheritance, and child custody. The duality of legal systems in some countries, where both civil and customary law exist side by side, hinders the implementation of international human rights instruments like CEDAW. This is because these instruments are civil law instruments, which cannot be codified into customary law. Furthermore, where customary law is practiced in a way which marginalises or discriminates against women as equal citizens, it is highly unlikely that human rights principles such as the right to equality and the provisions of other international instruments will be considered.

- Differential Access to and Control over Resources: Productive, reproductive and community roles require the use of resources. In general, women and men have different levels of both: access to the resources needed for their work, and control over those resources.

- Access: the opportunity to make use of something. Control: the ability to define its use and impose that definition on others.

- Economic/Political/Social/Time/Resources: Resources can be economic: such as land or equipment; political: such as representation, leadership and legal structures; social: such as child care, family planning, education; and also time — a critical but often scarce resource.

- Female Genital Mutilation (FGM): Female genital mutilation comprises all procedures that involve partial or total removal of the external female genitalia, or other injury to the female genital organs for non-medical reasons. The practice is mostly carried out by traditional circumcisers, who often play other central roles in communities, such as attending childbirths. Increasingly, however, FGM is being performed by medically trained personnel. FGM is recognized internationally as a violation of the human rights of girls and women. It reflects deep-rooted inequality between the sexes, and constitutes an extreme form of discrimination against women. It is nearly always carried out on minors and is a violation of the rights of children. The practice also violates a person's rights to health, security and physical integrity, the right to be free from torture and cruel, inhuman or degrading treatment, and the right to life when the procedure results in death.

- Female Infanticide: Killing of a girl child within weeks of her birth.

- Femicide: The killings of women and girls because of their gender. The causes and risk factors of this type of violence are linked to gender inequality, discrimination, and economic disempowerment and are the result of a systematic disregard for women's human rights. It occurs in an environment where everyday acts of violence are accepted and impunity is facilitated by the government's refusal to deal with the problems.

- Feminisation of migration: The growing participation of women in migration. Women now move around more independently and no longer in relation to their family position or under a man's authority (roughly 48 per cent of all migrants are women).

- Feminisation of Poverty: The majority of the 1.5 billion people living on 1 dollar a day or less are women. In addition, the gap between women and men caught in the cycle of poverty has continued to widen in the past decade, a phenomenon commonly referred to as "the feminization of poverty". Worldwide, women earn on average slightly more than 50 per cent of what men earn.

- Women living in poverty are often denied access to critical resources such as credit, land and inheritance. Their labour goes unrewarded and unrecognized. Their health care and nutritional needs are not given priority, they lack sufficient access to education and support services, and their participation in decision-making at home and in the community are minimal. Caught in the cycle of poverty, women lack access to resources and services to change their situation. Although in general, women are not always poorer than men, because of the weaker and contingent basis of their entitlements, they are generally more vulnerable and, once poor, have less options in terms of escape. This suggests that interventions to address women's poverty require a different set of policy responses.

- Feminism: Feminism is a movement for social, cultural, political and economic equality of women and men. It is a campaign against gender inequalities and it strives for equal rights for women. Feminism can be also defined as the right to enough information available to every single woman so that she can make a choice to live a life which is not discriminatory and which works within the principles of social, cultural, political and economic equality and independence. Feminism can be also defined as a global phenomenon which addresses various issues related to women across the world in a specific manner as applicable to a particular culture or society. Though the issues related to feminism may differ for different societies and culture but they are broadly tied together with the underlying philosophy of achieving equality of gender in every sphere of life. So feminism cannot be tied to any narrow definitions based on a particular class, race or religion.

- Feminist Theory: An examination of women's positions in society, based on the belief that current positions are unequal and unjust, which also provides tactics and criteria for change.

- Gender Analytical Information: Gender analytical information is qualitative information on gender differences and inequalities. Gender analysis is about understanding culture, e.g. the patterns and norms of what men and women, boys and girls do and experience in relation to the issue being examined and addressed. Where patterns of gender difference and inequality are revealed in sex disaggregated data, gender analysis is the process of examining why the disparities are there, whether they are a matter for concern, and how they might be addressed.

- Gender-Sensitive Budgets: or 'women's budgets,' refers to a variety of processes and tools, which attempt to assess the impact of government budgets, mainly at national level, on different groups of men and women, through recognising the ways in which gender relations underpin society and the economy. Gender or women's budget initiatives are not separate budgets for women. They include analysis of gender targeted allocations (e.g. special programs targeting women); they disaggregate by gender the impact of mainstream expenditures across all sectors and services; and they review equal opportunities policies and allocations within government services.

- Gender in Development (GID): The GID or Gender in Development perspective emerged in the late 1980's as an alternative to the prevailing Women in Development or WID approach. Unlike WID, which focused on women only, and called for their integration into development as producers and workers, GID focuses on the interdependence of men and women in society and on the unequal relations of power between them. The GID approach aims for a development process that transforms gender relations in order to enable women to participate on an equal basis with men in determining their common future. The GID approach emphasises the importance of women's collective organisation for self empowerment.

- There are two very similar terms in current usage – Gender in Development (GID): Gender and Development (GAD). There is no substantive difference in the meaning of these 11 two terms, which may be used interchangeably.

However, UNDP favours the use of the GID formulation, as it has a slightly more "integrated" connotation. Of course, if gender perspectives were fully mainstreamed into development thinking and action, there would be no need for either designation, as it would be understood that gender inequality is a fact of socioeconomic life, and therefore must be addressed as integral to all development initiatives.

- Gender in Development (Condition and Position): Development projects generally aim to improve the condition of people's lives. From a gender and development perspective, a distinction is made between the day-to-day condition of women's lives and their position in society. In addition to the specific conditions which women share with men, differential access means women's position in relation to men must also be assessed when interventions are planned and implemented. Condition: This refers to the material state in which women and men live, and relates to their responsibilities and work. Improvements in women's and men's condition can be made by providing for example, safe water, credit, seeds, (i.e. practical gender needs). Position: Position refers to women's social and economic standing in society relative to men, for example, male/female disparities in wages and employment opportunities, unequal representation in the political process, unequal ownership of land and property, vulnerability to violence (i.e. strategic gender need/interests).

- Gender, Institutions and Development Data Base (GID-DB): The Gender, Institutions and Development Data Base (GID-DB) represents a new tool for researchers and policy makers to determine and analyse obstacles to women's economic development. It covers a total of 161 countries and comprises an array of 60 indicators on gender discrimination. The data base has been compiled from various sources and combines in a systematic and coherent fashion the current empirical evidence that exists on the socio-economic status of women. Its true innovation is the inclusion of institutional variables that range from intra-household behaviour to social norms. Information on cultural and traditional practices that impact on women's economic development is coded so as to measure the level of discrimination. Such a comprehensive overview of gender-related variables and the data base's specific focus on social institutions make the GID-DB unique, providing a tool-box for a wide range of analytical queries and allowing case-by-case adaptation to specific research or policy questions.

- Gender Equality as a Development Objective: "At the United Nation Fourth World Conference for Women held in Beijing 1995, both DAC members and their partner countries made commitments to gender equality and women's empowerment. The Beijing Declaration adopted at the Conference builds on the perspectives and strategies outlined at the previous United Nations conferences on education – Jomtien, (1990), environment-Rio(1992), human rightsVienna (1993), population – Cairo (1994), and social development – Copenhagen (1995), including the Convention on the Elimination of all Forms of Discrimination against

Women (CEDAW, 1979). It is based on the principles of human rights and social justice. It clearly recognises that gender equality and women's empowerment are essential for addressing the central concerns of poverty and insecurity, and for achieving sustainable, people centred development." The follow-up to the Fourth World Conference on Women in 1999 further recognizes that gender mainstreaming is a tool for effective policy-making at all levels and not a substitute for targeted, women-specific policies and programmes, equality legislation, national machineries for the advancement of women and the establishment of gender focal points.

- Gender-Neutral, Gender-Sensitive, and Gender Transformative: The primary objective behind gender mainstreaming is to design and implement development projects, programmes and policies that:
  1. Do not reinforce existing gender inequalities-Gender Neutral
  2. Attempt to redress existing gender inequalities-Gender Sensitive
  3. Transformative: attempt to redefine women and men's gender roles and relations Gender Positive

- Gender Pay Gap: The percentage difference between the median hourly earnings of men and women, excluding overtime payments. The causes of the gender pay gap are complex-key factors include: human capital differences: i.e. differences in educational levels and work experience; part-time working; travel patterns and occupational segregation. Other factors include: job grading practices, appraisal systems, and pay discrimination.

- Gender Policies: Gender policies are divided into three categories depending on the extent to which they recognize and address gender issues

- Gender-aware policies: Gender-aware-policies recognise that women as well as men are actors in development and that they are often constrained in a different way to 15 men. Their needs, interests and priorities may differ and at times conflict. Gender aware-policies can be sub-divided into two policy types:

- Gender-neutral policies approaches use the knowledge of gender differences in a given context to target and meet the practical needs of both women and men. Gender-neutral policies do not disturb existing gender relations.

- Gender-specific policies use the knowledge of gender differences in a given situation to respond to the practical gender needs of either women or men. These policies do not address the existing division of resources and responsibilities.

- Gender-blind policies: Policies that are gender-blind fail to distinguish between the different needs of women and men in their formulation and implementation. Thus, such policies are biased in favour of existing gender relations and therefore are likely to exclude women or exacerbate existing inequalities between women and men.

- Gender-redistributive policies aim to transform the existing distribution of resources and responsibilities in order to create a more equal relationship between women and men. Women and men may be targeted or one group alone may be targeted by the intervention. Gender-redistributive policies focus mainly on strategic gender interests, but can plan to meet practical gender needs in a way which have transformatory potential (provide a supportive environment for women's self empowerment).

- Gender-specific torture and ill treatment: Gender often has a considerable impact on the form of torture that takes place, the circumstances in which it occurs, its consequences, and the availability of and access to remedies for its victims. Rape, threat of rape, electro-shock to the genitals and strip searching of women detainees by male guards are frequently the forms that such gender-specific torture. In societies where a woman's sexuality is a reflection of family "honour", these forms of torture and ill treatment are rarely reported.

- Gender Training: A facilitated process of developing awareness and capacity on gender issues, to bring about personal or organizational change for gender equality. "The generic aim of gender training is to consciously introduce gender as a category of analysis (as opposed to description), to point to the differing needs and interests of women and men and their unequal representation, and to increase awareness and reduce the gender-bias which informs the actions of individuals and institutions." This kind of gender training commonly involves:

- Raising participants' awareness of the different and unequal roles and responsibilities of women and men in any particular context

- Looking at ways that development interventions affect, and are affected by, differences and inequalities between women and men

- Equipping participants with knowledge and skills to understand gender differences and inequalities in the context of their work, and to plan and implement policies, programmes and projects to promote gender equality.

- Gender Violence: Any act or threat by men or male-dominated institutions that inflicts physical, sexual, or psychological harm on a woman or girl because of their gender. There is, however, no single definition of gender violence accepted internationally and there is much debate over the breadth of inclusion. Commonly, the acts or threats of such included in the definition are rape, sexual harassment, wife-battering, sexual abuse of girls, dowry-related violence, and non-spousal violence within the home. Other definitions extend to marital rape, acts such as female genital mutilation, female infanticide, and sex-selective abortion. In addition, certain definitions include 'sexual exploitation' such as enforced prostitution, trafficking of women and girls, and pornography.

- Hegemonic masculinity: Socially and historically constructed idea of what men ought to be, in a way a structure that links power to masculinity.

- Honour Killings: Acts of violence, usually murder, committed by male family members against female family members who are perceived to have brought dishonour upon the family.

- Human Rights: Human rights are basic rights and freedoms that all people are entitled to regardless of nationality, sex, national or ethnic origin, race, religion, language, or other status. Human rights include civil and political rights, such as the right to life, liberty and freedom of expression; and social, cultural and economic rights including the right to participate in culture, the right to food, and the right to work and receive an education. Human rights are protected and upheld by international and national laws and treaties.

- Human Rights with a Gender Perspective: This perspective recognizes that differences in life experiences based on gender often results in social, economic, political, and other inequities for women and girls. This view, when applied to policy development and service delivery, promotes positive change in the lives of women and girls. For example, home-based English as a Second Language program would allow immigrant mothers who care for their children at home to learn English and function in their new surroundings.

- Hyper-masculinity: Exaggerated image of hegemonic masculinity, mainly in media. It overemphasises the ideals set out for men hence reinforcing them.

- Intersectionality: Tool for analysis, advocacy and policy development that addresses multiple discriminations and helps understand how different sets of identities impact on access to rights and opportunities. Intersectionality is an analytical tool for studying, understanding and responding to the ways in which gender intersects with other identities and how these 20 intersections contribute to unique experiences of oppression and privilege. It starts from the premise that people live multiple, layered identities derived from social relations, history and the operation of structures of power. Intersectional analysis aims to reveal multiple identities, exposing the different types of discrimination and disadvantage that occur as a consequence of the combination of identities. It aims to address the manner in which racism, patriarchy, class oppression and other systems of discrimination create inequalities that structure the relative positions of women. Intersectional analysis posits that we should not understand the combining of identities as additively increasing one's burden but instead as producing substantively distinct experiences. It is therefore an indispensable methodology for development and human rights work.

- Phallocentric: The term refers to the cultural and social organization of the world fostered by the patriarchy.

- Politics of Location: A research perspective which grew out of feminist methodology, primarily through the critiques of women of colour in both the global north and global south who viewed the majority of early gender research as stemming from a generic white, Northern, middle-class perspective. The politics of location suggests that personal backgrounds and experiences of researchers (whether chosen or imposed by society) have political and theoretical implications that must be articulated throughout the research process.

- Polygamy: A man marrying more than one wife or temporary wives leading to insecurity of women and facilitating the spread of HIV/AIDS. It is illegal in most countries but still persists.

- Prejudice: Is made up of unfavourable or discriminatory attitudes (not actions) towards persons of different categories. Racial, sexual and other types of discrimination can exist at the level of personal relations and individual behaviour as well as be institutionalised as legal or administrative policy.

- Productive Work: This is work that produces goods and services for exchange in the market place (for income). Some analysts, especially those working on questions of equality between men and women, include the production of items for consumption by the household under this definition, even though they never reach the market place, regarding this as consumption of a form of non-monetary income. Both men and women contribute to family income with various forms of productive work, although men predominate in productive work, especially at the higher echelons of remuneration. Historically, in most societies, changes in economic structure, and hence in the structure of productive activities, have led to changes in the sexual division of labour and gender relations.

- Reflexivity: A key component of feminist research, reflexivity is the process through which researchers seek to constantly reflect upon, and critically analyze the nature of the research process — choosing methods, conducting research, writing the research project, proposing outcomes and solutions, and research presentation. Feminist researchers also use reflexivity to analyze the gender relations underlying not only the research subject in question, but the way of conducting research in general. Feminist researchers will commonly use self-reflexivity in their own research, but will also partake in collaborative reflexive techniques (such as consciousness-raising) to deepen their analyses via the perspectives of other researchers and also their research participants.

- Reproductive Rights: Reproductive rights embrace certain human rights that are already recognized in national laws, international human rights documents and other consensus documents. These rights rest on the recognition of the basic right of all couples and individuals to decide freely and responsibly the number, spacing and timing of their children and to have the information and means to do so, and the right to attain the highest standard of sexual and reproductive health. It also includes their right to make decisions concerning reproduction free of discrimination, coercion and violence, as expressed in human rights documents. The promotion of the responsible exercise of these rights for all people should be the fundamental basis for government-and community-supported policies and programmes in the area of reproductive health, including family planning.

- Sexual Exploitation: Any abuse of position of vulnerability, differential power, or trust for sexual purposes; this includes profiting momentarily, socially, or politically from the sexual exploitation of another.

- Sexual harassment: Sexual harassment is a form of sexual violence. The term refers to onesided, unwanted and unwelcome behaviour where sexuality and/or varied cultural constructions of sexuality are used as the means to oppress and position people and to produce or maintain vulnerability among them. Sexual harassment is often divided into two types: quid pro quo harassment and hostile environment harassment. The difference between these two types of harassment is that in quid pro quo harassment sex is provided in exchange for things such as employment or educational benefit – job promotion or good grades, for example – or the avoidance of some detriment. Hostile environment harassment means sexual harassment that creates an intimidating, hostile or offensive environment generally for a whole group of people-such as women, young women, some ethnic group of women or some groups of men, to mention a few. The forms of sexual harassment are usually divided into three different types: (1) verbal: e.g. remarks about figure/looks, sexual and sexist jokes, verbal sexual advances, comments that implicate stereotypic and discriminative attitudes; (2) non-verbal and/or visual: e.g. staring at someone and whistling; and (3) physical: acts from unsolicited physical contact to assaults and rape.

- Social Justice: Fairness and equity as a right for all in the outcomes of development, through processes of social transformation.

- Stereotypes: a generalized set of traits and characteristics attributed to a specific ethnic, national, cultural or racial group which gives rise to false expectations that individual members of the group will conform to these traits.

- Strategic Gender Interests: Needs and interests identified by women that arise from their subordinate position in society. Strategic interests vary according to context, are related to gendered divisions of labour, power and control, and may include such issues as legal rights, domestic violence, equal wages, access to contraception, etc. Strategic gender interests question women's socially constructed role, demanding greater equality and a change in existing roles.

- Trafficking in Persons: It was only in November 2000 that an international definition of trafficking was agreed to, under the Protocol to Prevent, Suppress and Punish Trafficking in Persons, Especially Women and Children (Trafficking Protocol), supplementing the United Nations Convention Against Transnational Organized Crime, 2000 (also known as the Palermo Convention):

  (a) "Trafficking in persons" shall mean the recruitment, transportation, transfer, harbouring or receipt of persons, by means of the threat or use of force or other forms of coercion, of abduction, of fraud, of deception, of the abuse of power or of a position of vulnerability or of the giving or receiving of payments or benefits to achieve the consent of a person having control over another person, for the purpose of exploitation. Exploitation shall include, at the minimum, the exploitation of the prostitution of others or other forms of sexual exploitation, forced labour or services, slavery or practices similar to slavery, servitude or the removal of organs;

(b) The consent of a victim of trafficking in persons to the intended exploitation set forth in subparagraph (a) of this article shall be irrelevant where any of the means set forth in subparagraph (a) have been used;

(c) The recruitment, transportation, transfer, harbouring or receipt of a child for the purpose of exploitation shall be considered "trafficking in persons" even if this does not involve any of means set forth in subparagraph (a) of this article;

(d) "Child" shall mean any person under eighteen years of age.

- Transformatory Potential: A gender analysis guided by this approach, applying the analytical framework to development programming, uses the interwoven framework of concepts to assess 30 the transformatory potential of a given set of options--which ones are most likely to ensure women get equal access to the resources they need to maximise their productive and reproductive contributions to their households and societies.

- Triple role/ multiple burden: These terms refer to the fact that women tend to work longer and more fragmented days than men as they are usually involved in three different gender roles – reproductive, productive and community work.

- Universal Declaration of Human Rights: The Universal Declaration of Human Rights (UDHR) is a milestone document in the history of human rights. Drafted by representatives with different legal and cultural backgrounds from all regions of the world, the Declaration was proclaimed by the United Nations General Assembly in Paris on 10 December 1948 General Assembly resolution 217 A (III) as a common standard of achievements for all peoples and all nations. It sets out, for the first time, fundamental human rights to be universally protected.

- Universal Declaration of Human Rights: The Universal Declaration of Human Rights (UDHR) is a milestone document in the history of human rights. Drafted by representatives with different legal and cultural backgrounds from all regions of the world, the Declaration was proclaimed by the United Nations General Assembly in Paris on 10 December 1948 General Assembly resolution 217 A (III) as a common standard of achievements for all peoples and all nations. It sets out, for the first time, fundamental human rights to be universally protected.

- Violence against Women (VAW): Article 1 of the UN Declaration on the Elimination of Violence against Women, proclaimed by the UN General

Assembly in its resolution 48/104 of 20 December 1993, defines the term "violence against women" as: "Any act of gender-based violence that results in, or is likely to result in physical, sexual or psychological harm or suffering to women , including threats of such acts , coercion or arbitrary deprivation of liberty, whether occurring in public or in private life. Three contexts of violence are differentiated in Article 2: Family, community and state. The forms shall be understood to encompass, but not be limited to, the following:

a) Physical, sexual and psychological violence occurring in the family: wife-battering, sexual abuse of female children in the household, dowry-related violence, marital rape, and female genital mutilation and other traditional practices harmful to women, nonspousal violence and violence related to exploitation.

b) Physical, sexual and psychological violence occurring within the general community: rape, sexual abuse, sexual harassment and intimidation at work and education institutions, trafficking in women and forced prostitution.

c) Physical, sexual and psychological violence perpetrated or condoned by the State, wherever it occurs.

- Violence (Domestic): A pattern of abusive and threatening behaviours that may include physical, emotional, economic and sexual violence as well as intimidation, isolation and 31 coercion. The purpose of domestic violence is to establish and exert power and control over another; men most often use it against their intimate partners, such as current or former spouses, girlfriends, or dating partners. Forms of domestic violence can include physical violence, sexual violence, economic control, and psychological assault (including threats of violence and physical harm, attacks against property or pets and other acts of intimidation, emotional abuse, isolation, and use of the children as a means of control). Because they occur in intimate relationships, many kinds of abuse are often not recognized as violence. In many places throughout the world, marital rape is still not viewed as sexual assault because a husband is deemed to have a right of sexual access to his wife. Stalking, as well, has only recently been recognized as a form of violence and a severe threat to the victim.

- Women's Studies: By focusing on the extent to which traditional questions, theories and analyses have failed to take gender into account, Women's Studies (as a field) adopts scholarly and critical perspective toward the experiences of women.

The objectives of Women's Studies include:
- Finding out about women by raising new questions and accepting women's perceptions and experiences as real and significant;

- Correcting misconceptions about women and identifying ways in which traditional methodologies may distort our knowledge;
- Theorizing about the place of women in society and appropriate strategies for change;
- Examining the diversity of women's experiences and the ways in which class, race, sexual orientation and other variables intersect with gender.

Although studying women is its starting point, by uncovering the ways in which social and cultural assumptions and structures are shaped by gender, Women's Studies also studies men and the world around us.

It is interdisciplinary, integrating insights from many different experiences and perspectives. Drawing from scholarly work within nearly every academic discipline as well as from the work of "grassroots" feminism, Women's Studies has its own distinctive, evolving theories and methodologies.

- Gender dynamics refers to the relationships and interactions between and among boys, girls, women, and men. Gender dynamics are informed by socio-cultural ideas about gender and the power relationships that define them. Depending upon how they are manifested, gender dynamics can reinforce or challenge existing norms.
- Parity in education refers to equivalent percentages of males and females in an education system (relative to the population per age group). Parity is essential but not sufficient for achieving gender equality.
- Constructive men's engagement: An approach to achieving gender equality that consciously and constructively includes men as clients, participants, supportive partners, and agents of change.
- Gender-accommodating: when project design, implementation, and evaluation approaches adjust to or compensate for gender differences, norms, and inequities by being sensitive to the different roles and identities of men and women, but in ways which do not change the status quo.
- Gender responsive: Being aware of how gender identities and roles influence the opportunities of men and women in society and designing activities and policies that are structured and operate to demonstrate a commitment to gender equality. This mean ensuring that women are among the participants and beneficiaries, whether as the extension agents hired, the farmers reached, or the scientists trained. It also means ensuring that both men and women have the appropriate training and skills to understand and support women farmers, extension agents, employees, and entrepreneurs.
- Gender indicators: An indicator is a measure. Gender indicators (or gender-related or gender-sensitive indicators) measure changes in specific conditions of men and women or on the level of disparity between them. The indicator may be constructed to show either an absolute measure.
- Gender targets: Targets establish how projects measure success. Gender targets are goals that are expected to be reached either to improve women's

and women's conditions relative to an earlier level OR to improve their situations relative to each other.

- Women's (and Girl's) Empowerment: A social process which enhances women's and girls' capacity to act independently (self-determination), control assets, and make choices and decisions about all aspects of one's life. Through women's empowerment unequal power relations are transformed, and women gain greater equality with men. At the government level this includes the extension of all fundamental social, economic, and political rights to women. At the individual level, this includes processes by which women gain confidence to express and defend their rights, and greater self-esteem and control over their own lives. The participation and acceptance of men in changing their own roles and supporting change among women is essential for achieving women's empowerment.

- Gender presentation: how a person dresses, looks, and acts; the presentation of one's sense of gender through behaviour and dress.

- Intra-household Resource Distribution: The dynamics of how different resources that are generated within or which come into the household, are accessed and controlled by its members.

- National Machineries for Women: Agencies with a mandate for the advancement of women established within and by governments for integrating gender concerns in development policy and planning.

- Gender Diversity: refers to the extent to which a person's gender identity, role, or expression differs from the cultural norms prescribed for people of a particular sex. This term is becoming more popular as a way to describe people without reference to a particular cultural norm, in a manner that is more affirming and potentially less stigmatizing than gender nonconformity.

- Gender Expression: An individual's presentation, including physical appearance, clothing choice and accessories, and behaviour that communicates aspects of gender or gender role. Gender expression may or may not conform to a person's gender identity.

- Gender Dysphoria: refers to discomfort or distress that is associated with a discrepancy between a person's gender identity and that person's sex assigned at birth (and the associated gender role and/or primary and secondary sex characteristics).

- Gender Non-Conforming is an adjective and umbrella term to describe individuals whose gender expression, gender identity, or gender role differs from gender norms associated with their assigned birth sex. Subpopulations of the TGNC community can develop specialized language to represent their experience and culture, such as the term "masculine of centre" that is used in communities of colour to describe a GNC identity.

- Gender Impact Analysis: Gender Impact Analysis is a systematic approach for assessing and understanding the different impacts of development on women and men because of their different gender roles.
- Gender Diagnosis: is the organization of data to highlight key gender problems, underlying causes of problems for men and women, and the relationship between problems and causes.
- Gender Strategy: Clear operational strategies, which will be used to achieve stated objectives, must identify the incentives, budget, staff, training, and organizational strategies to achieve stated objectives.
- Gender Monitoring And Evaluation: Flexible planning requires gender monitoring and evaluation to enable adjustment to experience and to establish accountability of commitment to achieve gender-specific priorities.
- Sex assignment is the initial categorization of an infant as male or female.
- Sexual Orientation A component of identity that includes a person's sexual and emotional attraction to another person and the behaviour that may result from this attraction. An individual's sexual orientation may be lesbian, gay, heterosexual, bisexual, queer, pansexual, or asexual. A person may be attracted to men, women, both, neither, gender queer, androgynous or have other gender identities. Sexual orientation is distinct from sex, gender identity, gender role and gender expression.

Δ Gender and Land Rights

Why are land rights important...?

Land rights are essentially claims which are legally and socially recognised and enforceable by an external legitimate authority, be it a village level institution or an authority of the State.

Land rights can be in the form of~

- Ownership: for example, private property, generally the most secure right
- Usufruct: for example, customary rights to common property resources
- Tenancy rights Land rights can stem from~
- Inheritance
- Community membership
- Transfers by the State (land reform)
- Tenancy agreements
- Purchase or land-leasing from the market (limited for women)
- Encroachment on public land (more limited and contentious)

Lack of access to land in rural India is of fundamental importance as it is linked to the incidence of poverty: "Aside from its value as a productive factor, land ownership confers collateral in credit markets, security in the event of natural hazards or life contingencies, and social status. Those who control land tend to exert a disproportionate influence over other rural institutions, including labour and credit markets"

Land reforms are seen as a critical means of redistributing land and transferring some rights to the poor and socially excluded groups. In India, land was made a state subject in 1935 and land reform was the responsibility of individual state governments. Although the degree of implementation of land reforms varies significantly from state to state, legislation in post-Independence India has been of essentially three types:

- Abolition of intermediary tenures during the 1950s, namely the colonial zamindari system, bringing substantial gains to many farmers at relatively low political cost but somewhat high financial cost (heavy compensation paid to zamindars).

- Regulation of the size of holdings through ceiling-surplus redistribution and/ or land consolidation: the former achieved little in terms of redistribution because of loopholes in the law which allowed landlords to retain control over landholdings if they could bribe the village patwari to register holdings in the name of deceased or fictitious persons (benami transactions). Land consolidation, although a popular strategy for empowering small and marginal farmers, has tended to benefit those with larger landholdings.

- Set dement and regulation of tenancy has also generally been weak and in manycases led to a worsening of tenure security. According to some estimates tenancy reforms have led to the loss of access (to land) by the rural poor to around 30 per cent of the total operated area). In states where tenancy reform has been successful, as in West Bengal, it has been due to the political system which supported legislative change, particularly through the PRIs and awareness campaigns (land settlement camps) which sought to enhance the bargaining power of tenants.

Given the inherent structural limitations of land reforms it is not surprising that women's land rights were never a policy concern till recently. Women in poor households may have benefited in terms of overall food security, but redistribution was always based on the household as a unit and men as household heads (except in the rare cases where widows or divorcees were eligible, see Jacobs 1998). As one official from the Ministry of Agriculture pointed out to Bina Agarwal, an internationally recognised authority on women's land rights: "You want women to have land?! What do you want to do? To destroy the family?" (cited in Agarwal 1994: 53).

Recognising the need for women to have land. Tenure, the Sixth FYP (1980-85) mandated that all redistributed land should be under joint title, but this has never been easy to implement. The Eighth FYP (1989-95) went a step further by calling for a change in inheritance laws to accommodate women's rights but gave few specifics. It called upon state governments to allot 40 per cent of ceiling-surplus or State redistributed land in the name of women alone, with the remainder to be held jointly (Mearns1999: 26). The Ninth FYP and the National Perspective Plan for Women: 1988-2000 made several substantive recommendations for closing the gender gap in access to land, but there is still little sustained focus on the question of women and land (Agarwal 2003: 186).

Δ Why are land rights for women important?

Essentially there are three types of arguments underlying the importance of land rights for women (Agarwal 1994): welfare, efficiency and empowerment.

- Security against poverty/welfare: To the extent that women's livelihood strategies are land-based, direct access to land for women means greater control over agricultural income, not just for their own well-being, but more importantly, for their children, since it is well-known that children's nutritional status is more closely related to their mother's income rather than the father's. Land owned by women enables them to be recognised as 'farmers' and provides them with indirect advantages such as access to extension services, collateral for credit or as an asset which can be mortgaged or sold during a crisis. For widows and the elderly, owning land could improve welfare not just directly, but also by enhancing their entitlement to family welfare (Agarwal2003: 194).

- Efficiency implications: Women are often the sole or de facto head of the household (e.g. when men have migrated) and independent land tenure provides them with production incentives – to adopt improved agricultural technology and practices, for example. However, such initiatives need to be supported by agricultural extension services and access to other inputs like credit – the organisation of women into Self Help Groups (SHGs) has been critical in this respect. While production inefficiency associated with tenure insecurity continues to be one of the important rationales for land reform, there is also emerging evidence, albeit contested, which suggests that women might use resources more efficiently than men in a given context.

- Empowerment: Land titles provide women with status/identity and strengthen their bargaining power in the family and community, though the extent to which women are able to challenge male dominance is contingent on several other contextual factors (e.g. her 'fallback' position, access to community social networks, ability to manage land). Where social movements such as the Bodhgaya movement in Bihar or NGOs like the Deccan Development Society in Andhra Pradesh have enabled poor rural women, particularly dafits, gain access to land, the women have vividly voiced their perceptions of change in their social status: "Having land in women's name has made an enormous difference-learning to take on land means taking on more power and wisdom. Once we got land, our eyes opened," claimed dafit women who had purchased land with the help of the DDS.

Δ Obstacles to women inheriting, accessing and managing land

Property relations are essentially social relations, that is, between people rather than between people and 'things'. In the context of gender inequalities, all over the world and to the extent that they have access to power and authority, it is men who wield greater claims to property. Even where laws have been modified to give women some inheritance rights there are a number of social factors which prevent them from exercising their (legal) claims and being able to effectively manage their property / landholdings, such as:

- Unequal personal laws which favour sons or male heirs. Despite some changes in inheritance rights in favour of women, both Hindu and Moslem personal law generally treat agricultural land differently from other property. Moreover, women's rights in tenancy land depend on state laws and here there are striking regional differences. In the three southern states of Andhra Pradesh, Tamil Nadu and Karnataka as well as Maharashtra, under amendments to the Hindu Succession Act (1956), women can inherit agricultural land, whether owned (by including daughters as coparceners in joint family property) or under tenancy, while Kerala has abolished joint family property altogether.

Social bias arising from:

Resistance to women inheriting land: Women are often accused of being witches, particularly in. Adivasi communities, as a means to not only torture them, but also to reclaim land and other economic assets.

Vulnerability of women: Women 'voluntarily' forego their claims in the face of violent threats from male relatives, or because they see their brothers as potential protectors from violence and ill-treatment inflicted by their in-laws.

- Administrative bias: Local officials are often unwilling to record women's inherited shares in land.
- Inability to manage/cultivate land:
- Prevalence of patrilocal manage patterns which means that most daughters leave their natal villages and find it difficult to manage land given by their parents.
- Restrictions on women's mobility, particularly in upper caste and Muslim families in north-western India (UttarPradesh).
- Dependency on male labour for ploughing, a commonly accepted norm all over the country, arising from women's alleged 'impurity', it is a social taboo for women to touch the plough.
- Gender bias in extension services. Limited access/control to cash or credit.

Given the constraints in granting women individual or joint land titles, many gender progressive development organisations have mobilised rural women (small and marginal farmers, landless and destitute) to collectively lease land (usually through their SHGs, simgamsor mandals), with or without joint cultivation and management as a prerequisite. Apart from DDS, other organisations that have facilitated such efforts include the Gorakhpur

Environmental Action Group (GEAG) and Society for Promoting People's Participation in Ecosystem Management (SOPPECOM)2 in Pune. In both these cases land has been leased from large farmers in the village for a fIxed period of time, with landless women collectively investing in resources using LEISA OOW external input sustainable agriculture) techniques towards sustainable agriculture promoted by the NGOs. Such efforts are parallel to ongoing legal struggles for access to land whether through consolidation for small and marginal farmers or the numerous campaigns against land alienation by the State of village commons and 'wastelands', in favour of industrial holdings.

- However, women's rights tend to get subsumed in these larger arenas and the need for separate spaces for women's participation and articulation of gender inequalities in property rights is important.

Δ Gender-sensitive stakeholder analysis

In order to ensure that women's as well as men's needs, priorities and constraints are recognised and addressed and influence the development agenda, all processes of policy development and project design should involve:

- participatory consultation with women as well as men in beneficiary groups
- women as well as men in decision-making at all levels
- gender equality advocates (men as well as women) devising ways of opening up spaces to ensure women's active involvement in consultation and decision-making.

This means finding ways to ensure that:

- women's groups are actively involved in consultation and decision-making processes
- The range of women's views and needs is adequately represented. Different women (and men) have different needs on the basis of class, ethnicity, age and family composition, and other factors. Urban, middle class women do not necessarily accurately represent the views and priorities of poor, rural women
- the usual processes of stakeholder analysis (drawing up a table of stakeholders; assessing the importance of each stakeholder and their relative power and influence; and identifying risks and assumptions that will affect project design) include:
- women and men as separate stakeholder groups
- where appropriate, different stakeholder groups amongst women (and men)
- clarity regarding stakeholder groups which include both women and men

- consultancy teams, working groups, management teams and implementation teams include women as well as men
- gender equality advocates (in government, civil society and donor organisations) work in collaboration, thinking collectively and strategically about advocacy strategies.

Δ Women in decision-making: community level

Issues to address

Traditionally, women are often excluded from decision-making at the community level. A number of factors combine to bring this about. These include traditional attitudes concerning the role and status of women, and also aspects of women's own work burden, knowledge, skills and confidence. Poor women's confidence can be undermined by less exposure than poor men to the world outside their immediate home, and by limited language and literacy skills. Even when steps have been taken to include women in community level decision-making, too often women have been token representatives on community committees with a passive role and few real responsibilities. Problems for women can be compounded during negotiations with local authorities. Community based groups may have been able to achieve considerable levels of women's participation, but decision-making power may lie at higher levels of the local administration, where women are not so well represented. Community efforts are often frustrated by bureaucratic delays or unwilling staff at the local/municipal government level, and women community representatives can be particularly vulnerable because of their generally lower social status.

Δ Increasing women's involvement in community decision-making Gender analysis

Before taking action to involve women in community level decision-making, it is important to be fully aware of existing gender roles, structures and attitudes in relation to decision-making at the community level.

Planning to promote women's involvement

Action to promote women's involvement in community level decision-making should be devised on the basis of a clear understanding of existing gender roles, and on the basis of male and female community members' perspectives and priorities.

On this basis:

- appropriate ways of strengthening women's involvement in decision-making need to be specified in planning documents, included in implementation staff TORs (terms of reference) and supported with necessary funding.

- criteria for monitoring and evaluation of women's participation must also be established. Indicators of effectiveness should include qualitative as well as quantitative aspects of participation.

Activities to promote women's involvement:

- Practical measures to promote women's involvement in decision-making include the following: Community consultation processes
- practical measures are needed to ensure that project information reaches women, that they are able to attend meetings and that meetings provide a forum in which they can actively participate
- women themselves will often have insights on the best way to work around male dominated power structures
- open discussions involving men and women may facilitate women's participation but specific measures may also be needed to overcome the deference or muting of women's views in front of men
- particularly in large communities, it may be necessary to follow up large meetings with smaller planning groups, including key women representatives, where women's roles, responsibilities, priorities and constraints can be elaborated in more detail
- given the limitations on poor women's time, considerable outreach work and flexibility is required about when and where to meet. One approach has been to arrange meetings in situ at, for example, water supply sites or clinics
- working with existing women's NGOs or community organisations is a way to involve women directly. However, such organisations tend to be monopolised by more affluent women with more free time, and may exclude poorer sections of the community
- women's organisations are not necessarily "gender-sensitive", in the sense that they may have limited understanding of ideas concerning gender mainstreaming and gender equality. It may be useful to take steps to strengthen the gender sensitivity of CBOs and networks.

Activities to gain the support of men

- early consultation with men, particularly community leaders, and attempts to promote positive attitudes towards women's active participation, are important. Where women are involved in separate activities or training, the potential advantages should be explained, and/or complementary or parallel activities organised for men
- men's negative attitudes to women's increased involvement have often shifted once the benefits to the community, households, and women themselves have been demonstrated.

Promoting women's active role in community level decision-making

- women's involvement in selecting candidates is likely to result in a higher and more dynamic level of women's participation
- the quality of women's participation in committees, as well as the quantity, needs to be improved. For women who are unused to assuming positions of authority, considerable groundwork may be needed to develop the self confidence and assertiveness skills necessary for dealing with village authorities. Women representatives may need special training, in leadership skills, confidence building and communication. Similar training should be offered to men to avoid alienation.

Links with local authorities

Local women's needs are often addressed most effectively by building gender-sensitive partnerships between community representatives and local authorities. This involves:

- Tupporting and training community representatives to negotiate effectively for gender sensitive services
- Training staff in municipal authorities to increase their understanding of gender issues, needs and rights, as well as their responsibility for delivering gender-aware responses
- Developing activities to increase information to marginalised groups, including women, about the services and resources they can expect, e.g. service charters setting out standards of provision.

Action to promote gender equality:

Gender mainstreaming is a strategy to promote the goal of gender equality and the empowerment of women.

Gender equality does not simply or necessarily mean equal numbers of men and women or boys and girls in all activities, nor does it necessarily mean treating men and women or boys and girls exactly the same.

It signifies an aspiration to work towards a society in which neither women nor men suffer from poverty in its many forms, and in which women and men are able to live equally fulfilling lives. It means recognising that men and women often have different needs and priorities, face different constraints, have different aspirations and contribute to development in different ways. It means recognising the right of women to define the objectives of development for themselves.

Δ   Outline gender equality action framework

Choice of action to promote gender equality should be made on the basis of clear gender analytical information and sex disaggregated data, and on the basis of women's own priorities and concerns. It is wholly inappropriate for development organisations to devise actions to promote gender equality and women's empowerment without having gone through these two steps.

This outline gender equality action framework is designed to assist in planning discussions. In all sectors and contexts, possible action to promote gender equality broadly falls into the listed categories. Agreed actions to promote gender equality should be included in policy and planning documents, and backed up with staffing, resources and indicators of change.

Type of Action	Issues to consider
ORGANISATIONAL LEVEL	
Information systems and research	-collating and commissioning targeted gender analytical research
-establishing sex disaggregated information systems	
Building the capacity of staff in management, policy-making and implementation agencies	-developing staff gender-related skills, knowledge and commitment through e.g. training workshops, consultancy support, provision of guidelines
Promoting gender equality in policy-making, management and implementation agencies	-development of procedures to promote equality in recruitment and career development
-identifying and addressing gender-related issues in the organisational culture	
Solidarity and networking	-activities to link together individuals and groups working for gender equality
BENEFICIARY LEVEL	
Addressing women's and men's practical needs	-recognising and addressing practical needs/problems identified by and particular to either women or men e.g. developing domestic labour saving devices for women
Promoting equality of access and benefit	-promoting greater gender equality in relation to resources, services, opportunities and benefits e.g. increasing women's access to previously male dominated employment opportunities
Increasing equality in decision-making	-promoting women's and men's equal participation in community level decision-making institutions and in community representation

Addressing the ideology of gender inequality	-working with beneficiary groups to reflect on gender norms, traditions and values e.g. participatory community workshops on HIV/AIDS
-addressing inappropriate gender stereotypes e.g. reviewing school text books for inappropriate gender stereotyping	

Δ Gender-sensitive indicators

Gender-sensitive Indicators allow measurement of benefit to women and men. Depending on the policy/project, this might include:

- The impact/effectiveness of activities targeted to address women's or men's practical gender needs i.e. new skills, knowledge, resources, opportunities or services in the context of their existing gender roles
- The impact/effectiveness of activities designed to increase gender equality of opportunity, influence or benefit. e.g. targeted actions to increase women's role in decision-making; opening up new opportunities for women/men in non-traditional skill areas
- The impact/effectiveness of activities designed to develop gender awareness and skills amongst policy-making, management and implementation staff
- The impact/effectiveness of activities to promote greater gender equality within the staffing and organisational culture of development organisations e.g. the impact of affirmative action policies.

There is no standard or agreed-upon method for measuring women's empowerment. Aspects of empowerment can be reflected in numbers (such as an increase in numbers of women in positions of power), but above all, empowerment concerns women's perceptions of their own lives and experiences. To measure qualitative aspects of empowerment, it is important that it is clearly defined. Most definitions stress two main areas:

- A personal change in consciousness involving a movement towards control, self-confidence and the right to make decisions and determine choices
- Organisation aimed at social and political change.

The greater the degree of existing gender inequality, the more subtle changes are likely to be. It is important in this context for indicators to recognise the significance of modest gains and breakthroughs.

Δ How do they measure?

Gender-sensitive indicators need to capture quantitative and qualitative aspects of change

Δ Quantitative indicators

Quantitative indicators refer to the numbers and percentages of women and men or organisations involved in or affected by any particular group or activity. Quantitative indicators draw on the sex disaggregated data systems and records that have been examined during processes of policy or project planning. The availability of quantitative baseline data means that indicators usually include some element of target setting.

For example:
- Women form at least 33% of water committee members by the end of Year 2.
- At least 50% of network members have developed a gender policy by the end of Year 3.
- Equality in girls and boys access to primary education by 2005.
- 25% increase in number of female police officers by 2005, from a baseline of x%.

Monitoring information should be available through routine data systems and records.

Δ Qualitative indicators

Qualitative information refers to perceptions and experiences. Qualitative information is vitally important. It is not enough to know that women are participating in an activity: the quality of their participation and experience, whether in community level meetings, primary school classes or as users of public services, is all-important.

Qualitative indicators (as well as quantitative indicators relating to visible change at the community level) should be developed in conjunction with beneficiary groups. In project documents it is legitimate to use in a phrase like "quantitative and qualitative indicators to be developed with beneficiary groups in first six months of the project". This creates the space to develop indicators in conjunction with beneficiary groups once they have fully understood the nature of the project. (What changes would they like to see? What will the change look like? How can it be measured?). This process should take place using qualitative methods such as focus group discussions and informal interviews.

It is only possible to set targets for qualitative change if baseline data is available. This requires baseline surveys: it is highly unlikely that appropriate baseline data will be available from secondary sources.

Where baseline data is available on experiences and perceptions, targets for qualitative change can be set.

For example:
- At least 50% of women participating in water committees report active involvement in management and decision-making by the end of Year 2 (from a baseline of 10% at the start of the project)
- At least 70% of women respond positively to evaluation of police handling of their case in targeted police stations by the end of Year 3 (from a baseline of 5% average at the start of the project).

Where baseline data is not available, or is not easily aggregated into numbers and percentages, it is necessary to resort to general statements of improvement.

For example:
- Significant improvement in staff knowledge, skills and attitudes on mainstreaming gender equality in participating organisations by the end of Year 3 (where each organisation starts with markedly different levels)
- Significant increase in quantity and improvement in quality of media reporting on gender violence.

Information on qualitative indicators should be collected through evaluation surveys. Depending on the indicator, these might be questionnaire surveys reviewing perceptions and experiences of agreed indicators, or participatory methods such as focus group discussions and case studies.

Δ PIMs marker on removal of gender discrimination checklist

To achieve a significant score for "removal of gender discrimination", projects should meet the following criteria:
- equality between women and men is explicitly promoted in activity documentation (i.e. the project explicitly aims to promote benefit for women as well as men)
- gender analysis has been carried out either separately or as an integral part of standard procedures (i.e. gender analytical information should be included, at least, in the social annex and the social appraisal section of the project memorandum)
- gender analysis has been incorporated into activity design such that the activity meets a number (at least 4) of the following criteria:
- gender-sensitive strategies and implementation plans are incorporated and reflected in the activity budget

- specific means have been designed to help overcome identified barriers to women's full participation in the activity
- specific means have been included to help ensure equitable control by women and men over activity output
- gender-sensitive indicators including impact indicators have been or will be developed for monitoring and evaluation
- gender skills are used in design and will be used in monitoring and evaluation
- Gender-sensitive consultation is carried out at all levels and stages (i.e. women as well as men have participated in project design and will continue to be involved in implementation and management).

Δ Gender focal staff

Evaluations of gender mainstreaming repeatedly and consistently conclude that effective gender mainstreaming in any context requires staff (not consultants), variously referred to as gender focal persons/change agents/gender "entrepreneurs"/gender advocates, to take responsibility for spearheading, supporting and sustaining gender work. The role of these staff is not to take full personal responsibility for gender work, but to act as catalysts supporting and promoting gender-related skills and approaches amongst professional colleagues. The evidence is overwhelming that unless there are staff with designated responsibility, responsibility for gender equality all too easily becomes "mainstreamed" out of existence.

Δ The role of gender focal staff

The role of staff with responsibility for promoting gender mainstreaming involves:
- thinking strategically about where efforts and available resources should be focused
- identifying and taking advantage of opportunities and "entry points" to mainstream gender
- seeking out allies (internally and externally)
- mobilising resources
- providing convincing justifications for the relevance of gender mainstreaming to the organisation and its activities
- facilitating the development and monitoring of gender policy and action plans
- developing and monitoring systems and procedures for mainstreaming gender
- Developing and supporting gender mainstreaming skills, knowledge and commitment with professional colleagues and partners i.e. through training, guidelines and support.

Δ Common constraints

The effectiveness of gender focal points, particularly those based in sectoral and regional ministries and project implementation teams, has often proved disappointing, at least partly because expectations of what they might achieve can be unrealistically high. Gender focal points commonly face the following constraints:

- lack of clarity about their roles and responsibilities
- lack of management support
- no additional time/resources allocated to their gender focal point role
- women staff members selected as focal points on the basis of their sex rather than their commitment to the issues
- Relatively junior staff members selected as focal points but lacking the authority and seniority to undertake this role effectively. The potential for introducing change from below in an organisation accustomed to hierarchical top down forms of decision-making is inevitably limited
- huge demands on their personal and professional initiative and resilience
- Many existing gender analysis methodologies and training packages are oriented to data gathering/analysis at the community/project level. A lot of this is not directly relevant to gender focal points operating at a policy level.

Δ Promoting effectiveness

Positive focal point experiences, associated with promoting tangible change and sustaining momentum, are strongly related to supportive management, scope and resources for developing and implementing policy and activities, and adequate support. Donors have an important role to play in facilitating the effectiveness of gender focal points both in partner organisations and in project implementation teams.

- Focal point TORs: terms of reference for the gender focal point should be clearly spelt out, and, if appropriate, developed in conjunction with senior managers and gender focal points themselves. TORs should realistically bear in mind the time and resources that will be available to individual focal points, and confirm the role of the gender focal point as a catalyst
- capacity-building: this could include training (in gender mainstreaming and advocacy skills), mentoring, links to professional networks, participation in workshops
- Professional and personal support: through backstopping support and involvement in networks.

Δ Gender policies

Introduction

An organisational mission statement/policy is a useful starting point for gender mainstreaming. Once gender equality is being effectively addressed in mainstream policy documents, a specific and separate gender mainstreaming policy may no longer be necessary.

Content

A gender mainstreaming policy usually includes:

Background information

Problem/situation analysis, focusing on beneficiary groups. What is the evidence for gender inequality? Why is it a problem? Relate this to your own organisational goals. Use appropriate sex disaggregated data and gender analytical information what is being done (generally) to address the issue of gender inequality. Existing/previous government/ NGO initiatives and approaches. Focus on:

- achievements
- challenges
- lessons learned
- focus on own organisation
- history in addressing the issue of gender inequality
- current work and responsibilities
- Achievements/challenges/lessons learned.
- Ways forward.

Δ Policy commitments

- Succinct statement of policy rationale (a statement of organisational vision and mission in relation to gender equality. Statements of principle and belief including words like "we believe" or "we recognise"). For example:

"We believe that women and girls are overrepresented amongst the poor, marginalised and oppressed, as a result of the unequal distribution of power and resources between women and men in all societies."

- Succinct statement of policy commitments in relation to specified areas of work (statements of action including words like "we will". It is possible and quite helpful to use a logical framework format for this). For example:

"We will provide appropriate training and support to all staff to ensure they have adequate awareness, knowledge and skills with which to concretely address gender issues in their work."

Strategy

A strategy is an action plan to put policy commitments into practice. In relation to all policy commitments, it is important to specify the following:

- activities
- indicators
- time frame
- designated responsibility
- Budget.

Δ Lessons learned

- Policy evaporation

  All too often, gender mainstreaming policies "evaporate" before implementation, and remain paper commitments only. Policies must include strategies/action plans with clear procedures and targets as well as designated roles and responsibilities for promotion, implementation, and monitoring.

  These must be based on a clear and realistic analysis and understanding of the organisation/department including its decision-making structures, incentive systems, planning routines and history with respect to gender equality.

- Focus on process and product

  The value of a gender mainstreaming policy lies at least as much in its formulation as in its existence. The formulation of a mainstreaming policy is a golden opportunity to involve as many staff and, where appropriate, stakeholder's external to the organisation as possible. This process promotes widespread "ownership" of the policy; enhances understanding and commitment to gender equality issues; ensures that the policy "fits" with the Organisational culture, structures and procedures; and substantially increases the chance that the policy will be implemented. In this context:

  - gender policies from other similar organisations can be used for ideas and inspiration, but should never be copied or used as blueprints
  - External consultants may have a useful role to play in facilitating a consultation and policy development process, but should never be recruited to write a mainstreaming policy.

Δ Practice what you preach

Gender equality in the workplace, and gender equality in-service delivery, are inextricably linked. Agency credibility in presenting a gender equality policy relating to service delivery is assisted if the policy is reflected in or includes measures to promote gender equality in internal staffing and practice.

Δ Gender Training
What is gender training?

In-service gender training emerged in the mid 1980s to "teach" development policy makers, planners and implementation staff to see and take account of the differential impact of development interventions on men and women. This kind of gender training commonly involves:

- raising participants' awareness of the different and unequal roles and responsibilities of women and men in any particular context
- looking at ways that development interventions affect, and are affected by, differences and inequalities between women and men
- Equipping participants with knowledge and skills to understand gender differences and inequalities in the context of their work, and to plan and implement policies, programmes and projects to promote gender equality.

It has been, and remains, quite common for development agencies and governments to develop short (often one or two day) gender-awareness planning courses designed to be applicable to all staff within the organisation. More recently, many development agencies are moving away from this "one size fits all" approach to gender training onto a more tailored approach.

Δ "Best practice" in gender training: the context

Gender training is most effective when used as part of a broader strategy for influencing the climate of opinion within an organisation for promoting gender equitable practice. Equally, the importance attached to gender training by the organisation as a whole influences how seriously training is taken by course participants. Participants who expect some sort of follow up activity, and whose supervisors support and promote gender equitable practice, are more likely to transfer what they have learned to their working practice. Activities complementary to gender training will vary with circumstance. Part of the role of staff with responsibility for promoting attention to gender equality is to identify appropriate entry points and opportunities. Possibilities might include:

- follow up discussion and feedback workshops
- participatory gender policy development with clear, measurable and achievable objectives
- ensuring staff have back-up access to gender expertise and to professional support
- inclusion of attention to gender equality issues in personnel appraisals
- forming internal gender networks and committees
- working with external advisory/consultative groups
- establishing earmarked funds for pilot initiatives

- activities to promote management support for gender mainstreaming
- active monitoring of gender policy implementation

Δ "Best practice" in gender training: the content

The GEM website includes information on how to go about planning a tailored gender training course and suggestions for gender training exercises. It is important to bear in mind in all contexts that gender training works most effectively when:

Δ It is learner centred

- All training should be based on an analysis of the participants and their needs. The more homogenous the group of participants, the more the training can be tailored to their specific needs, the more effective it will be.
- It uses participatory methods.
- Effective training uses participatory methods such as case studies, brainstorming, and problem solving to allow participants to actively engage with the subject matter, and learn by doing. Choice of methods will depend on the topic, the group, the trainer and practical factors. It is important to use country, culturally and sectorally specific case material directly relevant to the circumstances in which participants live and work. The participants' own policies, projects, experiences, observations and deliberations should be the principal materials for discussion.

Δ It introduces skills as well as awareness

- Effective training is based on an understanding of the participants own job responsibilities, an understanding of where they fit in their organisational structure and an understanding of their organisational systems and procedures. It should help participants to identify and discuss their own opportunities and constraints to develop a gender equality perspective, and encourage the development (and follow up) of personal action plans.

The trainer has credibility with the participants

- The trainer needs to have knowledge, understanding and status appropriate to the group. In all circumstances trainers need to adopt a non-threatening approach allowing discussion and exploration of different viewpoints. It is often best for external consultants to work with internal gender staff in order to ensure the relevance of the training to the organisation.

Δ Training is followed up

- Competence development is a process not an event. Training needs to be followed up with discussion workshops, more tailored training and/or on-the job support.

Δ Pitfalls in gender training

The above conclusions on "best practice" in the context and content of gender training are well rehearsed, but all too often gender training fails to reach these standards. Whilst good gender training can promote a more positive climate of opinion to facilitate gender equitable work, poor gender training not only fails to promote gender equitable practice, it can provoke a backlash to hard-won progress. It can promote opposition to participation in any further gender training and/or an inappropriate sense of having "done gender". Resistance is part of the territory of gender training, and will be encountered by good gender trainers in good gender training courses, as well as by bad gender trainers in bad gender training courses. However, gender trainers bear responsibility for predicting and managing resistance constructively, and this requires their explicit attention to all of the above points on best practice in gender training content. Ineffective gender training cannot and should not simply be blamed on resistance.

Δ Too much gender training provokes resistance and/or is ineffectual because:

- it is formulaic
- it is dislocated from the needs of the group
- it says more about the trainer than the trainees: it is "too academic", it is "too feminist", it regurgitates what the trainer learnt on a training of trainer course.

Δ Commissioning gender training

In commissioning gender training, it is centrally important to be aware of best practice in both the context and content of gender training and to ensure, as far as possible, that this is followed. If you are commissioning gender training, it is quite likely that you will also be responsible for promoting gender mainstreaming in others ways. It is essential to consider ways in which the training will be reinforced and followed up.

Δ In terms of the content of gender training:
Work alongside external gender training consultants

- it is preferable for external gender training consultants to work alongside staff responsible for promoting attention to gender mainstreaming within the organisation in order to ensure the relevance of the training to the organisational culture, structures and procedures, and to ensure that the training complements and reinforces other mainstreaming initiatives

Δ Use a team of trainers rather than an individual trainer

Training is often conducted most effectively by teams rather than individuals. This is partly because gender training can be extremely challenging and tiring, and co-facilitators can give each other support and feedback. It is also because, in moving from "one size fits all" to training tailored to the needs of the participants, it is unlikely that one trainer will have all the knowledge and skills required. Co-training is also a way of building training capacity.

Δ Factors to consider in selecting trainers

- Gender trainers have different areas of expertise as well as different styles and approaches to training, i.e. they do not all do the same job in the same way. Find out all you can about the approach of different gender trainers from people who have experienced working with them. Think about what kind of expertise and approach would be appropriate to the needs of your participants, and discuss this with potential trainers. It is important to think about the credibility of the trainer/s with the group

- it is important for at least one trainer to come from the same area and ethnic group as the majority of the participants

- male gender trainers can stop gender being seen as a woman's issue, and promote the credibility of gender mainstreaming in mixed and/or largely male groups

- trainers with highly developed theoretical understanding of gender analysis may be desirable for highly educated, academic groups and policymaking groups, but less appropriate for groups more concerned with practical details of planning and implementation

- trainers with practical and applied experience of mainstreaming gender in particular sectors may be desirable for sector-specific groups

- Trainers with an overtly radical/feminist approach may be appropriate to groups already committed to mainstreaming gender equality and/or women's groups.

Δ Allow time and resources for needs analysis and planning

Training must be tailored to the needs and roles of the participants. Trainers must be allowed time and resources to conduct effective needs analysis, and to develop appropriate and tailored training materials.

Δ Promoting gender training capacity

There has been an enormous increase in demand for gender trainers in the last few years and, with the current increase in attention to gender mainstreaming in accordance with the Beijing Platform for Action commitments, this demand is likely to increase still further. In response to demand, there has been a proliferation in many countries of "gender trainers" and "gender training institutes". Whilst some of the gender training provided in this context is very good, in too many cases gender training capacity is weak and quality poor. It is important for donors to support and develop local gender training capacity as much as possible.

Quite a lot of training of gender trainer courses have trained participants in a standard gender training. There is often a case for developing and repeating a standard gender training course within a particular organisation (for example, when training a large number of staff playing a similar role within the same organisation), and a consequent need to train trainers in the use of that particular training package. It is important, however, to be quite clear about the purpose and the limitations of training trainers in one training package. It does not produce trainers able to devise and tailor gender courses to different institutional and participant needs, and trainers using a standard training package in a setting for which it was not designed will provide poor quality training.

The move towards tailored gender training is much more demanding on trainers. It requires trainers with gender-related knowledge and skills sufficiently wide ranging to meet the needs of potential course participants, and with the confidence and skills required to assess the learning needs of participants and develop and conduct training courses accordingly.

Effective gender training skills build up with experience as well as training. Training of effective gender trainers is not a one-off event. Donors can support the development of effective gender training through:

- Tailored training of trainer courses (moving away from the idea of "one size fits all" gender training). For example:
- training in gender training for sector based workers and consultants, focusing on gender analysis and gender equitable practice in particular sectors, for example, health work; policing; macro-economic policy etc.
- training of gender trainers in advocacy, lobbying and influencing techniques
    - training of gender trainers in institutional analysis and gender equitable practice in the workplace
    - Training of people with gender expertise/experience in training skills i.e. needs analysis, course planning, choice of methods, participatory monitoring techniques etc.
    - building the gender and training knowledge and skills base of trainers trained in a standard gender training package
    - training of activists/people active in the women's movement in gender analysis, through Masters courses and academic short courses
    - Facilitating access to/sharing of/publication of gender training materials.

Δ Management Support

A constant theme in effective gender mainstreaming is the importance of both the commitment and leadership of senior management. Only senior management can properly oversee a cross-cutting theme which, by definition, intersects the various management structures of the organisation. Senior management provides signals about the relative priority assigned to various issues through making demands on staff for analysis, information and updates on progress. When such demands are not made, and when staff is not held accountable for action on issues of equality, there is little incentive for action.

Equally, senior management support for those spear heading gender equality work is a key to success. Mainstreaming gender equality is a highly sensitive issue and often meets with staff opposition. The authority and support of senior management is important in enabling gender staff to continue in the face of resistance. Gender mainstreaming is often promoted on the basis of considerable trial and error and experimentation. Management support plays an important role in providing gender staff with the necessary space to try out different and at times controversial activities.

Δ Tips for improving the effectiveness of training programmes for women Farmers

- Women dislike long lectures and can more effectively learn while doing; hence the programme should be practical.
- Women prefer discussing problems that they currently face.
- Women prefer training programmes at locations closer to home.
- Training on crop practices should be between 2-7 days long and paced in a way that complements the agricultural calendar rather than interfering with it. Training should be provided ahead of the land preparation/sowing operation and at the stage of crop maturity.
- The desirable time for meetings is in the afternoon, when women are relatively free.
- Illiteracy is very high among rural women and long notes are of limited use.
- Audio-visual material should be used to the maximum extent.
- Use of local dialects is important in focussing attention.
- Involvement of women training and extension officers would increase effectiveness of the programme.
- For effective programmes and participation of women, it is necessary to have as much homogeneity as possible in the groups chosen for training or extension meetings.
- Special efforts should be made to promote interaction and provide opportunities for practical work.
- Women's training must be planned according to their preferences, learning needs and abilities.

Δ Operational Constraints

Most of the problems faced in the implementation of projects and programmes by the DoA are generic in all states. These problems include:

- restriction on expenditure on fuel for office vehicles, which affect the number of visits that can be made for monitoring performance of projects or programmes.

- Use of vehicles purchased specifically for women development projects and programmes are used for many other programmes of the government causing delays and cancellation of project activities.

- fixed touring allowance which is not enough to meet the travel requirements of project staff resulting in compromises in project activities. Public transport is unreliable and usually not available to villages, especially interior villages. Furthermore, women extension staff are not comfortable using them especially after dark.

- Delays in the release of funds thereby affecting project activities in the first few months of the financial year. Consequently, activities are not tailored to the agricultural calendar as they should be and some activities are not undertaken because farmers are busy in their fields or because the activity is not useful at that particular time of the growing cycle.

- Large jurisdictions for staff as vacancies are not filled in time. Scattered coverage of villages over a wide area causing a considerable amount of time spent on commuting to the project area.

- Special staff employed for implementing women programmes drawn for implementing other programmes

- lack of time for follow-up activities due to too many training camps and targets to be met by many other programmes being implemented by the DoA.

- Women programmes being treated as unimportant leading to low staff morale of those Employed in their implementation.

- Low honorarium for facilitators and their contractual term of appointment leading to low staff confidence.

- Women development projects and programmes in effect seen as a separate entity from mainstream DoA operations.

- Lack of co-ordination between different agencies involved even when mechanisms for coordination in the form of committees have been constituted.

- Lack of flexibility for the staff implementing the project to re-appropriate funds and to make changes in planned activities to meet the objectives of the programme thus affecting the ability to quickly respond to the needs of clients.

The constraints identified above need to be addressed for the full realisation of the benefits of woman farmers programmes. Some of these constraints have been addressed in the operational design of the Agricultural Technology Management Agency (ATMA) model of extension under implementation in 28 selected districts in the country.

Δ Gender friendly tools

Operation	Field Traditional Technology	Improved Technology
preparation In hilly areas	Spade	Simple hand tools/power packs for seed bed preparation
Sowing/planting	Hand dropping, pushing seedling in mud	Improved multi-row drills for seedling/fertiliser application
		Rotary dibblers, jab planter Manual seed drill/seed cum fertiliser drill Animal and power operated seed cum fertiliser drill
		6 row rice transplanter
Fertiliser application	Manual broadcasting	Fertiliser broadcaster
Weeding/hoeing/thinning	Khurpi, kudali, spade	Manual weeder, wheel hoe, garden rake
Irrigation	Flooding	Sprinkler and drip irrigation system
Spraying/dusting	Hand sprayer/duster without safety devices	Hand operated/foot operated sprayer with safety devices
Harvesting	Sickle	Serrated sickle, self propelled reaper of I m size
Threshing	Manual beating Bullock treading	Mechanical power thresher
		Pedal operated thresher Strippers
Preparation of soil and filling of polybags	Hand	Power operated hammer hills Hand scoop for filling poly bags
Watering	Bucket and mugs	Watering can, Wheel barrow for bringing water
Pruning/budding/grafting	Local knives, shears	Improved horticultural tools
Pit digging	Khurpi, spade	Augers and post hold diggers
Seed treatment	Hand mixing of seed with chemicals	Manually operated seed treatment drums

Cleaning/grading	Manual using cleaning basket/wire screens	Hand/pedal operated cleaners for grains/seeds
		Manual power operated cleaners, Winnowers
		Power operated graders
Drying	Sun drying Drying in cribs	Solar dryers Oil fired batch dryers
		Power operated dryers
		Agricultural waste fired dryers
Storage	Local storage structure made of clay, straw, bamboo, etc	Metallic storage structures
Milling	Hand mortar and pestle Foot operated Dhenki Hand operated stone grinders	Pedal operated grain mill Power operated grain mill, dal mill Wet grinder
Parboiling	Using cemented tank, metallic kettles and traditional methods of sun drying and milling	Parboiling equipment
Puffing and flaking	Using earthen pot, karhi, stirrer, broom, basket, oven, dhenki for milling	Rice puffing machine Flaking machine
Shelling, de-husking, decortication	Manual Knife.spike	Manual and power operated de-hullers Decorticators
		Hand shellers
Oil expression	Ghani	Portable power ghani
		Table oil expellers
		Screw expellers
Peeling, pulping, slicing, polishing	Knives, spikes etc	Manual and power operated peeler and slicer
Grinding of spices	Hand operated pounder	Mills/pulverisers (power operated)
Cream separation from milk, khoa making	Hand operated churns, manual methods	Power operated churns, khoa machines
Pappad making	Rolling pins	Hand/pedal and power operated presses
Leaf cup plate making	Manual	Power operated machine

Source: Gajendra Singh, Gyanendra Singh and Nachiket Kotwaliwale (1998) Mechanisation and Agro-Processing technologies for Women in Agriculture, Paper presented at the AIT-GASAT Asia Conference (August 4-7, 1998).

Gender impact assessment

Gender impact assessment has been defined as an ex ante evaluation, analysis or assessment of a law, policy or programme that makes it possible to identify, in a preventative way, the likelihood of a given decision having negative consequences for the state of equality between women and men.

The central question of the gender impact assessment is: Does a law, policy or programme reduce, maintain or increase the gender inequalities between women and men?

The European Commission defines gender impact assessment as follows:

"Gender impact assessment is the process of comparing and assessing, according to gender relevant criteria, the current situation and trend with the expected development resulting from the introduction of the proposed policy."

"Gender impact assessment is the estimation of the different effects (positive, negative or neutral) of any policy or activity implemented to specific items in terms of gender equality."

The assessment involves a dual-pronged approach: the current gender-related position in relation to the policy under consideration, and the projected impacts on women and men once the policy has been implemented. It is important that the assessment is structured, i.e. systematic, analytical and documented. The final aim of the gender impact assessment is to improve the design and the planning of the policy under consideration, in order to prevent a negative impact on gender equality and to strengthen gender equality through better designed, transformative legislation and policies.

A primary objective is to adapt the policy to make sure that any discriminatory effects are either removed or mitigated. Beyond avoiding negative effects, a GIA can also be used in a more transformative way as a tool for defining gender equality objectives and formulating the policy so as to proactively promote gender equality.

The final aim of the gender impact assessment is to improve the design and the planning of the policy under consideration, in order to prevent a negative impact on gender equality and to strengthen gender equality through better designed, transformative legislation and policies. A primary objective is to adapt the policy to make sure that any discriminatory effects are either removed or mitigated. Beyond avoiding negative effects, a gender impact assessment can also be used in a more transformative way as a tool for defining gender equality objectives and formulating the policy so as to proactively promote gender equality.

When is gender impact assessment needed?

The Council of the European Union, in its conclusions from 2006, noted that despite some progress toward gender mainstreaming in Member States, gender impact assessment still needs to either be put in place or reinforced. The Council urged in particular to improve and strengthen the development and regular use of gender impact assessment when drafting:

- Legislation
- Policies
- Programmes
- Projects

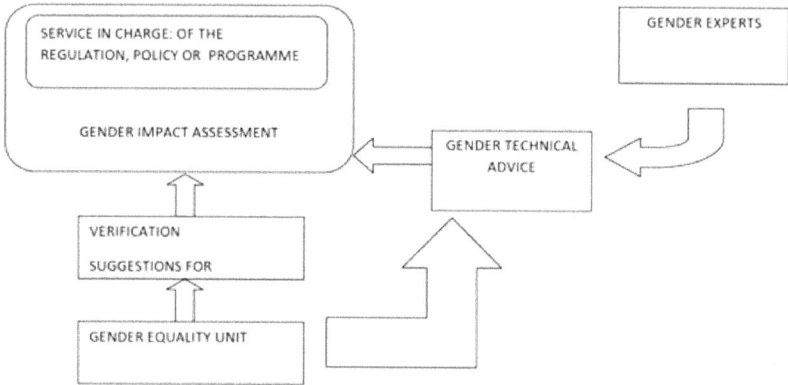

Diagram: gender impact assessment

Steps to carry out gender impact assessment

Some defined steps are needed to carry out a gender impact assessment, even though the number of steps defined may vary from one context to another, depending on the approach taken. However, the stages of the GIA process are always the same.

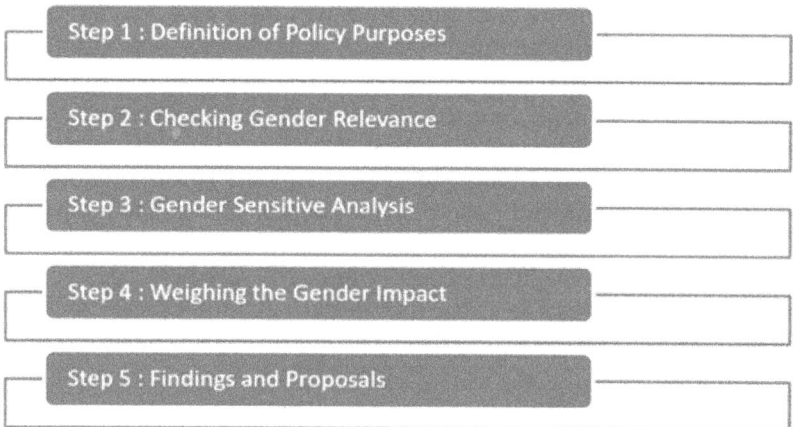

Gender Audit

Gender Audit

Women and men from various social groups are positioned differently in society-in the workplace and family-and have different responsibilities and needs. Due to this, women and men have different experiences, knowledge, talents and needs. Therefore, different programmers/schemes are designed to address this imbalance. Public services like education, health, transportation and welfare are utilized differently by women and men; as a result budgets impacts men and women differently.

UNESCO defines gender audit as management and planning tool. Gender Audit evaluates the gender responsiveness of an organization's culture and how well the organization is integrating the gender perspective into its work. The audit outcome aims to assist the organization to become more gender responsive. It is also a kind of impact evaluation that assesses or measure the impact of interventions on gender equality and women's empowerment. Traditionally audits are commonly associated with accounting audits in the financial world, used to certify that the finances and administration are legitimate, with established rules and regulations correctly followed. In the 1980s, quality management audit was introduced in companies to promote improvement of company performance. Following the same principles, social audits have been developed in a range of organizations to measure the extent to which they live up to "the shared values and objectives" to which they are committed. Social audit is a framework which allows an organization to build and develop a process whereby it can account for its social performance, report on that performance and draw up an action plan to improve on that performance. Gender Audit is one such form of social audit. Therefore, it can be summed up by stating that Gender Audit is a guided process to examine "what has been done" and "what has not been done to meet the Governments stated objectives of gender equality.

The definition of Gender Audit as adapted from various international definitions would be as below:

A Gender Audit is a tool to assess the accountability to and extent of Gender Budgeting accomplished-the integration of gender concerns into policies, strategies, programmes and schemes of all types. Its aim is to see whether the policy, guidelines, practices, systems, procedures and budgets are being used in the most effective way to deliver the Government's commitments to gender equality.

Characteristics of Gender Audit

1. Gender Audit takes into account whether internal practices and related support systems for gender mainstreaming are effective and reinforce each other.
2. Monitors and assesses the relative progress made in promoting gender sensitivity
3. Identifies critical gaps and challenges

4. Recommends ways of addressing these challenges and suggests new and more effective strategies to overcome gaps

5. Documents good practices towards the achievement of gender equality

A Gender Audit provides the tools to audit the processes which have been undertaken by any Department or organizations under the various policies, schemes and programmes. It examines whether and at what level Gender Budgeting initiatives have been adopted in the planning, implementation and review of the policies and budgetary allocations of the Department. The results of Gender Audit can provide learning and guidelines for Gender Budgeting to revisit policies, programmes and schemes as well as implementing mechanisms to ensure Gender mainstreaming. The Gender Audit will help the institution to identify gaps between their own gender equality policy and practice, and the way their programmes impact on gender equality.

ILO (A manual for gender audit facilitators: The ILO participatory gender audit methodology, 2007) defines audit as a tool to assess the extent of Gender mainstreaming accomplished – it helps to assess the differential impact of projects and policies on women and men. Its aim is to see whether the policy, programme guidelines, practices, systems, procedures and budgets are being used in the most effective way to deliver the Governments commitments to gender equality.

It states that a gender audit:

*   Considers whether internal practices and related support systems for gender mainstreaming are effective and reinforce each other and whether they are being followed ;

*   Monitors and assesses the relative progress made in gender mainstreaming ;

*   Establishes a baseline ;

*   Identifies critical gaps and challenges ;

*   Recommends ways of addressing them and suggests new and more effective strategies ;

*   Documents good practices towards the achievement of gender equality.

Here in a gender audit is essentially a "social audit", and belongs to the category of "quality audits", which distinguishes it from traditional "financial audits". It considers whether internal practices and related support systems for gender mainstreaming are effective and reinforce each other and whether they are being followed. The belief is that a gender audit enhances the collective capacity of the organization to examine its activities from a gender perspective and identify strengths and weaknesses in promoting gender equality issues. It monitors and assesses the relative progress made in gender mainstreaming and helps to build organizational ownership for gender equality initiatives and sharpens organizational learning on gender through a process of:

*   Team building

- Information sharing
- Reflection on gender

(Source: ILO Participatory Gender Audit)

Difference between Gender Audit, Gender Assessment and Gender Analysis Study

Gender Analysis Study	Gender Assessment	Gender Audit
Conducted during program design or planning	Conducted during project implementation or as part of the final evaluation after a project has been completed	Conducted during project implementation Assesses the internal and institutional context in which the program operates
Analyzes the gender context in which the program is operating	Assesses the program (both technical processes and result)	Evaluates gender integration as it relates to: policies, staff capacity, tools, training and resources, organizational culture and workplace issues
Helps programmers understand gender roles, responsibilities, statuses and inequalities so they can use that information to design, implement, monitor and evaluate programs	Assesses the extent to which the program integrates gender issues into activities and institutional systems. Assesses program results and effects on male and female participants Conducted via similar research processes as gender analysis, but includes a comprehensive review of existing activities and initiatives	Conducted via: desk review, questionnaire and interviews and focus groups with staff. Should ideally be carried out by external consultants. Makes recommendations about how to better mainstream gender within organization and among staff.
Conducted via: desk review, key informant interviews and interviews/ focus groups with beneficiaries	Includes recommendations based on results and lessons	
Includes recommendations on how to integrate Gender considerations into the program	Can include components of a gender analysis study, particularly if one has not already been done	

A gender audit enhances the collective capacity of the organization to examine its activities from a gender perspective and identify strengths and weaknesses in promoting gender equality issues. It monitors and assesses the relative progress made in gender mainstreaming and helps to build organizational ownership for gender equality initiatives and sharpens organizational learning on gender. To do this, it:

- considers whether internal practices and related support systems for gender mainstreaming are effective and reinforce each other and whether they are being followed;
- monitors and assesses the relative progress made in gender mainstreaming;
- establishes a baseline;
- identifies critical gaps and challenges;
- recommends ways of addressing them and-suggests new and more effective strategies;
- documents good practices towards the achievement of gender equality.

A participatory gender audit is a tool and a process based on a participatory methodology to promote organizational learning on how to practically and effectively mainstream gender in policies, programmes and structures and assess the extent to which policies have been institutionalized at the level of the:

- Organization
- Work unit
- Individual
  The participatory gender audits' objectives are to:
- Generate understanding of the extent to which gender mainstreaming has been internalized and acted upon by staff;
- Assess the extent of gender mainstreaming in terms of the development and delivery of gender-sensitive products and services;
- Identify and share information on mechanisms, practices and attitudes that have made a positive contribution to mainstreaming gender in an organization;
- Assess the level of resources allocated and spent on gender mainstreaming and gender activities;
- Examine the extent to which human resources policies are gender-sensitive;
- Examine the staff sex balance at different levels of an organization;
- Set up the initial baseline of performance on gender mainstreaming in an organization with a view to introducing an ongoing process of benchmarking to measure progress in promoting gender equality.

The ILO has developed a participatory gender audit methodology which has generated great international interest among constituents, donor's agencies, training organizations and academic institutions after more than ten years of implementation.

Chapter 2

Gender Resource Centre

Introduction

Rural Women form the most important productive work force in the economy of majority of the developing nations including India. Agriculture, the single largest production endeavour in India, contributing 25% of GDP, is increasingly becoming a Female Activity. Agriculture sector employs 4/5th of all economically active women in the country. 48% of India's self-employed farmers are women. There are 75 million women engaged in dairying as against 15 million men and 20 million in animal husbandry as compared to 1.5 million men.

Beyond the conventional market-oriented narrower definition of 'productive workers', almost all women in rural India today can be considered as 'farmers' in some sense, working as agricultural labour, unpaid workers in the family farm enterprise, or combination of the two. Moreover, several farm activities traditionally carried out by men are also being undertaken by women as men are pulled away into higher paying employment. Thus, Rural India is witnessing a process which could be described as Feminization of Agriculture.

More than simply supplying labour, women possess detailed knowledge of agriculture and use of plant and plant product for food, medicine and animal feed. Women today are central to the selection, breeding, cultivation, preparation & harvest of food crops. Apart from their pivotal role in cultivation of staple crops, they are primarily responsible for the production of secondary crops such as pulses and vegetables which are often the only source of nutrition available to their families. Women farmers also often possess unique knowledge about fish farming and handle most of the work associated with it.

Background & Justification

Efforts have been initiated in the recent past both by Governmental and Non-Governmental Organizations to incorporate Gender Issues into the Development Agenda to ensure women's full and equitable participation in agricultural development programmes. However, statistics still indicate that these efforts have not been sufficient enough to bridge gender inequalities. FAO study conducted recently found that women in developing countries contributed about 80% towards food production but received only 2 to 10% of the extension support.

The National Agriculture Policy (NAP) has also highlighted incorporation of gender issues in the agriculture development agenda recognizing women's role as farmers and producers of crops and live stocks; as users of technology; as active agents in marketing, processing and storage of food and as agricultural labourer. The policy states that high priority should be accorded to recognition and mainstreaming of women's role in agriculture. Appropriate structural, functional and institutional measures are proposed to be initiated to empower women and build their capacities and improve their access to inputs such as land, credit and agricultural technologies. Therefore, both for consideration of sustainability and equity, it is essential that Agricultural Research and Extension is more women centred reflecting the role of women as farmer.

Despite significant contribution of women in the production process, an all pervasive bias of development planners in treating them primarily as consumers of social services rather than producers, kept them away from the development programmes in agriculture and allied sectors. As men and women have different roles and needs and face gender-specific constraints, women may not automatically benefit from development activities, but may remain excluded.

Census 1991 had registered that only 22.3% of adult female population of India are workers, which may be a gross under statement since much of the work that women do, other than in the domestic and care sectors, is not recorded in the work participation format of the census. In fact, a pilot Time Use Survey conducted by the Central Statistical Organization came out with the starting revelations that 51% of the work of the women which qualify for inclusion in GDP are not recognized and remain unpaid.

The studies carried out so far in the field of agriculture indicate that despite the key role of women in crop husbandry, animal husbandry, fisheries, forestry and post-harvest technology, those in charge of formulating packages of technologies, services and public policies for rural areas have often tended to neglect the productive role of women. Consequently, the developments of technologies, specifically tailored to women-specific occupations and the involvement of women in technology development and transfer have received inadequate attention from both scientific and administrative Departments of Government.

Women have traditionally been discriminated in their access to productive resources and have been denied ownership of land, cattle, trees, harvest and shelter. They have even been discriminated in access to credit and marketing facilities for their economic activities. It is, thus, essential to develop strategies and mechanism to improve women's access to agricultural support services. Various Ministries/Departments are working towards the goal but have often tended to function in compartmentalized manner. These fragmented efforts are not sufficient to bring about the desirable impact.

Developing gender desegregated data and gender budgeting are the other key activities which can help bridge gender inequalities.

Gender Resource Centre (GRC)

The GRC is contemplated to be a focal point for convergence of all issues related to 'Gender in Agriculture'. The GRC would ensure that policies in agriculture reflect the national commitment to Empowerment of Women.

The GRC would not only undertake and support training, research and advocacy on gender issues in Agriculture & Natural Resources Management (NRM) but would forge effective functional linkages with other related departments, agencies and institutions.

Building on the experiences of women specific programme on agriculture that have been in operation, the GRC would identify critical gaps and help bridge these gaps along two paths: one- through 'strategy of mainstreaming' and other through 'strategy of agenda setting'. Mainstreaming simply means that women must be given opportunity to fully participate and benefit from all types of existing agricultural programmes. Agenda setting means that women farmers may be provided with structural and material resources so that they can participate and benefit at par with male farmers in setting their development agenda.

Roles and Responsibilities of Gender Resource Centre

1. To collect, analyse and document information (both from primary and secondary sources) on women in agriculture.
2. To act as a comprehensive data base & a clearing house to women related policies/issues in agriculture and allied sectors including NRM.
3. To monitor and assess the Gender impact of various on going programmes of agriculture and allied sector including NRM of Ministry of Agriculture and make recommendation on appropriate improvements in their design/ strategy.
4. To assess the gender impact of agricultural technologies and Research Project on 'women in agriculture', identify/assess the agronomic based drudgery prone activities of women and suggest ways to make these technologies/tools gender friendly.
5. Identify & float macro/micro level studies to identify the needs, requirements, potential and constraints faced by women in agriculture sector especially in the areas of technological development, access to inputs, credit and other productive resources, marketing intervention etc.
6. To review the existing laws and other Government decisions/measures relating to basic production resources such as land, water forest and to examine women's access and control over these basic resources and recommend necessary changes to protect women farmers right to livelihood.
7. To document, scientifically validate and disseminate traditional/ indigenous knowledge of women in agriculture and allied sector.

8. To forge effective functional linkages with various departments, agencies and institutions including non governmental organizations and farm women groups; document and disseminate lessons and experiences from on going initiatives taken by these agencies/institutions in sustainable agriculture and NRM.

9. To collaborate with Agri. Research institutions to identify technologies/ crops/ processes in which women farmers have a comparative advantage and develop a strategy for systematic capacity building on these issues.

10. To undertake preparation of suitable training modules on gender issues in agriculture and NRM which include gender sensitization modules for policy planners and development managers.

11. To view the existing policies related to land, water, forests with respect to their impact on women farmers and suggest remedial measures to bring about structural changes, if required.

12. To promote 'action research' on critical issues including women's access to land, water, common property resources, impact of macro economic changes on women farmer and implications of legal and regulatory framework on vulnerable groups such as land less farmers, tribal farm women & those affected by natural calamities.

13. To organize national level interactions between policy makers/ administrators and women farmers to share concern, issues and perspectives and evolve concrete policy recommendations. For the present, the work of GRC will be undertaken by the Women Cell of Directorate of Extension which will be restructured as needed.

Chapter 3

Women Empowerment

Introduction

"Empowerment is demonstrated by the quality of people's participation in the decisions and processes affecting their lives. In theory, empowerment and participation should be different sides of the same coin. In practice, what passes for popular participation in development and relief work is not in any way empowering to the poorest and most disadvantaged people in society."

- Empowerment of women is a socio-political ideal envisioned in relation to the wider framework of women's rights. It is a process that leads women to realise their full potential, their rights to have access to opportunities, resources and choices with the freedom of decision making both within and outside home. Empowerment would be achieved only when advancement in the conditions of women is accompanied by their ability to influence the direction of social change gained through equal opportunities in economic, social and political spheres of life.

- The Constitution of India conveys a powerful mandate for equality and rights of women in its Preamble, Fundamental Rights, and Duties and also provides for specific provisions for affirmative actions. India is also a signatory to a number of UN Conventions, primarily Convention on Elimination of all Forms of Discrimination against Women (CEDAW), Beijing Platform for Action and Convention on Rights of the Child where the commitment of the nation to protect and empower its women and girls is quite pronounced. The recent endorsement by India, of the ambitious 2030 Sustainable Development Goals (SDGs) will further change the course of development by addressing the key challenges such as poverty, inequality, and violence against women, which is critical for the global success of the goals as well.

- Notwithstanding the Constitutional mandate, the discourse on women's empowerment has been gradually evolving over the last few decades, wherein paradigm shifts have occurred – from seeing women as mere recipients of welfare benefits to mainstreaming gender concerns and engaging them in the development process of the country.

- Nearly a decade and half has passed since the National Policy for the Empowerment of Women (NPEW), 2001 was formulated, which laid down a comprehensive progressive policy for the advancement, development and empowerment of women with appropriate policy prescriptions and strategies. Since then, significant strides in global technology and information systems have placed the Indian economy on a trajectory of higher growth, impacting the general populace and women in particular, in unique and different ways.

- These changes have brought forth fresh opportunities and possibilities for women's empowerment while at the same time presenting new and emerging challenges which along with persisting socio-economic problems continue to hinder gender equality and holistic empowerment of women. Further, the deep-rooted patriarchal social constructs position women in disempowered social and economic hierarchies that impede in realization of their rights.

- Several paradoxical trends have been observed in the past few years. The growing acknowledgement of gender rights and equality is juxtaposed against increase in reporting of various forms of violence against women such as rape, trafficking, dowry etc.; expansion of new work opportunities for women alongside continued weak bargaining power in the labour market; increasing number of educated, aspiring career women entering the work place, while large sections of women are still in the low paid informal sector. Similarly, feminisation of agriculture and growing number of women farmers raises the larger issue of gender entitlements to land and assets ownership; growing state of art medical facilities has to be analysed against high Maternal Mortality Ratio (MMR), Infant Mortality Rate (IMR), malnutrition and anaemia and lack of geriatric care and support; growing urbanisation and resultant migration of women in relation to the availability of safe spaces and social security net for vulnerable women. Though climate change has significant gendered impact, women have been sidelined in debates about managing climate change and environmental resources. The misuse of rapid advances in Information Technology (IT) and telephony has resulted in new and varied forms of sexual abuse such as cyber crimes and harassment of women through mobiles and internet. The regulatory frameworks are not in pace with the growth in technology as yet.

- Investment in basic social infrastructure and services such as education, health, food security and nutrition, social protection, legal empowerment and poverty alleviation programs, will continue to be of paramount importance. However, the new millennium and the dynamics of a rapidly changing global and national scenario have ushered in new facets of development and growth giving rise to complex socio-economic and cultural challenges for women in a society with deep rooted cultural and social beliefs about gender roles.

- The current status of women with respect to human development parameters, legal rights for women to life and freedom from violence, economic and social discrimination and their rights to equality and equity shows that a lot still remains to be done. It is necessary therefore, to reinforce the rights-based approach for creating an enabling environment in which women can enjoy their rights.

- In the coming years, India is expected to gain significantly from it's 'demographic dividend' as the share of it's potential productive workforce will increase in numbers as compared to the aging population of other countries. To what extent the country can seize this dividend and benefit from it will largely depend on how women exercise their rights and entitlements and contribute to the development process.

- There is need therefore to formulate a new policy that can guide the transformative shift required for making gender rights a reality, addressing women's issues in all its facets, capturing emerging challenges and ultimately positioning women as equal partners of sustained development progress that the country is experiencing presently.

- The re-scripting of women's empowerment has been envisaged as a socially inclusive right based approach while reinforcing the rights and entitlements provided under the Constitution of India. The policy will enable sustainable socio economic, political empowerment of women to claim their rights and entitlements, control over resources and formulation of strategic choices in realisation of the principles of gender equality and justice.

- As a generic policy framework, the National Policy for Women, 2016, guides various sectors to issue more detailed policy documents that are sector specific. Sectors will be able to integrate the principles contained in this Policy Framework into their prevailing policy, strategy and program documents.

Distinguishing empowerment

- Empowerment has become a widely used word. In spheres as different as management and labour unions, health care and ecology, banking and education, one hears of empowerment taking place. The popular use of the word also means that it has been overextended and applied in circumstances that clearly do not involve much power acquisition beyond some symbolic activity or event.

- Empowerment in its emancipator meaning, is a serious word--one which brings up the question of personal agency rather than reliance on intermediaries, one that links action to needs, and one that results in making significant collective change. It is also a concept that does not merely concern personal identity but brings out a broader analysis of human rights and social justice.

- To gain a greater understanding of the concept, it might be helpful to look into its origins among popular movements. It emerged during the U. S . civil rights movements in the 1960s, after substantial work took place in civil disobedience and voter registration efforts to attain democratic rights for Afro-Americans. Displeased with the pace and scope of the changes, several black leaders (headed by Stokeley Carmichael) called for "black power," which they defined as: A call for black people in this country to unite, to recognize their heritage, to build a sense of community. It is a call for black people to begin to define their own goals, to link their own organizations, and to support those organizations (Carmichael and Hamilton, 1967, p. 44).

- Empowerment began to be applied within the women's movements in the mid 1970s. The similarities among oppressed groups are considerable because they face the common problem of limited willingness by those in control to see the seriousness of their condition and to work to solve it. Under the circumstances, the oppressed must themselves develop power for change to occur; power will not be given to them for the asking.

- Applied to gender issues, the discussion of empowerment brings women into the political sphere, both private and public. Its international use probably began with the appearance of the book by Sen and Grown, Development, Crisis, and Alternative Visions: Third World Women's Perspectives (1985), prepared for the Nairobi Conference at the end of the U.N. Decade for Women in 1985. In this book, a section on "Empowering Ourselves" clearly identifies the creation of women's organisations as central to the design and implementation of strategies for gender transformation.

- Women and men are placed in bipolar categories by numerous institutions in society. These institutions, through day-t o-d a y practices embedded in long-standing beliefs, construct male and female subjects who face strong forces to conform. Family practices, religious myths, the social division of labour, the sexual division of labour, marriage customs, the educational system, and civil laws combine to produce hierarchies, internalised beliefs, and expectations that are constraining but at the same time "naturalised" and thus seldom contested.

- In this context, empowerment is a process to change the distribution of power, both in interpersonal relations and in institutions throughout society. Traditionally the state has interpreted women's needs to suit its own preferences. The typical and enduring consideration that women have received from the state has been in their capacity of mothers and wives. Women therefore need to become their own advocates to address problems and situations affecting them that were previously ignored.

- Empowerment ultimately involves a political process to produce consciousness among policy makers about women and to create pressure to bring about societal change. There is an additional point to be made. Empowerment is a process which should centre on adult women for two central reasons: first, their adult lives have produced many experiences of subordination and thus they know this problem very well, although they have not labeled it as such and second, the transformation of these women is fundamental to breaking the integrational reproduction of patriarchal authority.

- 'Empowerment' and 'participation' became popular terms from the mid-1970s onwards as a response to the failure of top-down development planning to address the real needs of poor and marginalized people. However, there is no uniformly accepted definition of either concept-they have a variety of meanings depending on the institutional context and the nature of social actors involved. The concept of empowerment is closely related to participation whether in political processes (such as rural women's election to local panchayats) or participatory approaches to development involving rural women. But participation per se, which is essentially enabling access, does not necessarily lead to empowerment.

- At the root of the term empowerment is the notion of power. And power can be understood as operating in a number of different ways

- Power over: This kind of power involves an either/or relationship of domination/ subordination. Ultimately, it is based on socially sanctioned threats of violence and intimidation, it requires constant vigilance to maintain and it invites active and passive resistance (internalized oppression). This kind of power is predominantly wielded by men over other men, men over women and by dominant social, political, economic or cultural groups over those who are marginalized. It is thus an instrument of domination whose use can be seen in people's personal lives, their close relationships, their communities and beyond.

- Power to: This relates to having decision making authority, the power to solve problems for example, and can be creative and enabling.

- Power with: This involves people organising with a common purpose or common understanding to achieve collective goals.

- Power within: This refers to self-confidence, self-awareness and assertiveness. It defines the process through which individuals analyse how power operates in their lives and gain the confidence to act and influence this positively.

Defining empowerment

"Empowerment involves challenging the forms of oppression which compel millions of people to play a part in their society on terms which are inequitable, or in ways which deny their human rights" (Oxfam 1995 cited in Oxaal and Baden 1997).

"Empowerment is the process of challenging existing power relations and of gaining greater control over the sources of power" (Batliwala 1993).

"Empowerment requires transformation of structures of subordination through radical changes in law, property rights and other institutions that reinforce and perpetuate male domination" (Sen and Grown 1995, cited in Banerjee 1995).

"Empowerment is the process through which women can participate effectively in decisions that affect their lives-at the family, community and higher levels of the political system" (ISED 1991, cited in Banerjee 1995).

"Women's empowerment should lead to the liberation of men' from false value systems and ideologies of oppression. It should lead to a situation where each one can become a whole being regardless of gender, and use their fullest potential to construct a more humane society" (Akhtar 1992 cited in Batliwalla 1993:131).

"Empowerment is more than simply opening up access to decision-making; it must also include the processes that lead people to perceive themselves as able and entitled to occupy that decision .making space" (Rowlands 1995:102).

Characteristics of women empowerment

- Women empowerment is giving power to women. It is making women better off. It enables a greater degree of self-confidence and sense of independence among women.

- Women empowerment is a process of acquiring power for women in order to understand their rights and to perform her responsibilities towards oneself and others in a most effective way. It gives the capacity or power to resist discrimination imposed by the male dominated society.

- Women empowerment enables women to organize themselves increase their self-reliance and it provides greater autonomy.

- Women empowerment means women's control over material assets intellectual resources and ideology. it challenges traditional power equations and relations.

- Women empowerment abolishes all gender base discrimination in all institutions and structures of society. it ensures participation of women in policy and decision-making the process at domestic and public levels.

- Women empowerment means exposing the oppressive powers of existing gender social relations.

- Women empowerment makes women more powerful to face the challenges of life, to overcome to the disabilities, handicaps, and inequalities. it enables women to realize their full identity and powers in all spheres of life.

- Empowerment also means equal status to women. it provides greater access to knowledge and resources greater autonomy in decision.

- Women empowerment occurs within sociology, psychological, political cultural, familial and economic spheres and at various levels such as individual, group and community.

- Women empowerment is an ongoing dynamic process which enhances women's abilities to change the structure and ideologies that keep them subordinate. Women empowerment is a process of creating awareness and capacity building.

- In the process of empowerment women should consider their strengths and weakness opportunities and threats and move forward to unfold their own potential to achieve their goals through self-development, in our country empowering women through enterprise development has become an integral part of our development efforts due to three important advantage entrepreneurs, economic growth and social stability.

Objectives of women empowerment

i) Creating a conducive socio-cultural, economic and political environment to enable women enjoy de jure and de facto fundamental rights and realize their full potential.

ii) Mainstreaming gender in all-round development processes/ programmes/projects/ actions.

iii) A holistic and life-cycle approach to women's health for appropriate, affordable and quality health care.

iv) Improving and incentivizing access of women/ girls to universal and quality education.

v) Increasing and incentivising work force participation of women in the economy.

vi) Equal participation in the social, political and economic spheres including the institutions of governance and decision making.

vii) Transforming discriminatory societal attitudes, mindsets with community involvement and engagement of men and boys.

viii) Developing a gender sensitive legal-judicial system.

ix) Elimination of all forms of violence against women through strengthening of policies, legislations, programmes, institutions and community engagement.

x) Development and empowerment of women belonging to the vulnerable and marginalized groups.

xi) Building and strengthening stakeholder participation and partnerships for women empowerment.

xii) Strengthen monitoring, evaluation, audit and data systems to bridge gender gaps.

Priority areas for women empowerment
I. Health including food security and nutrition

i) Maternal and peri-natal mortality will remain a priority to bring down the high rates of MMR and IMR. The outreach and capacity of the frontline workers like ASHAs, ANMs, Anganwadi workers and the number of trained skilled home birth professionals with supportive supervision will be increased in remote and backward areas.

ii) A coordinated Referral Transport System (RTS) for safe deliveries as well as emergency obstetric care will be made available in all areas especially in difficult, remote and isolated areas and during the time of natural/environmental calamities.

iii) Focus on Pregnant and Lactating (P&L) women will be continued by organizing special health camps for the prevention and treatment of diseases affecting P & L women such as Anaemia, low BMI, under nutrition etc., and launch special drives to impart health and nutrition education.

iv) A gender transformative health strategy which recognises women's reproductive rights with shifts such as family planning focus from female sterilisation to male sterilisation will be developed and implemented.

v) Apart from maternal health, the focus of other health problems of women including communicable and non-communicable diseases like cancer, cardio vascular disease, HIV/AIDS will be given prioritised attention with appropriate strategies and interventions.

vi) Taking into account the vulnerable status of elderly women above 60 years of age who constitute 8.4% of the population, geriatric health care will require special attention in conformity with the National Policy on Senior Citizens 2011. Geriatric services including preventive, curative, and rehabilitative healthcare will be strengthened with appropriate government and public-private sector participation.

vii) Health interventions will aim at both physical and psychological well-being of women.

viii) The National Mental Health Policy (2014) recognises that women have a greater risk of mental disorders due to various reasons primarily due to discrimination, violence and abuse. A systematic approach to provide requisite screening, care and treatment especially at primary level will be made.

ix) Special emphasis will be given to the health care challenges of adolescents and investments in their health care. Adolescent sexual and reproductive health needs will get a focus of attention as a health issue in all health centres and hospitals.

x) There is a dearth of health care for women in their menopausal age. Though, the phase is generally dismissed as a natural occurrence, women are increasingly facing physical and emotional health problems like osteoporosis, cardiovascular diseases, depression etc. Suitable interventions will be initiated in this regard.

xi) Traditional knowledge of women including healing practices will be promoted for indigenous treatment systems in remote areas.

xii) Special attention will be given to the expansion of health insurance schemes such as RSBY and the same linked to programmes like ICDS, JSY, NRLM, NREGS, NULM benefitting women particularly the vulnerable and marginalised.

xiii) To improve the health services, complete, accurate, and timely gender based data as well as organizational data is urgently required. Investments in data infrastructure for linking gender based data collected by public and private organizations and individual researchers will be made to address the health reforms in other diseases apart from maternal health and to monitor the impact of e-initiatives.

xiv) As assisted reproduction pregnancies are more risky due to the high rate of multiple births and the risk of infection, healthcare coverage will be provided to the surrogate during the pregnancy, during post-pregnancy checkups and treatments that follow.

xv) Nutrition will be accorded utmost priority as women are at high risk for nutritional deficiencies in all the stages of their life cycle. To tackle the problem of malnutrition, focussed attention will be paid at every stage right from Ante-Natal Care (ANC) and Post Natal Care (PNC) for healthy foetal development to the needs of adolescent girls to the stage of elderly women. Interventions and services for addressing the intergenerational cycle of under-nutrition, with focus on nutritional care for the first 1000 days of the child after birth will be strengthened.

xvi) Improving the nutrition and health status of adolescent girls will be accorded special focus. Suitable strategies to end intra-household discrimination in nutritional matters with regard to girls and women will also be devised. Regular data on prevalence of nutritional deficiency in Children sex-wise (district wise data) and data on weight at birth (by sex) will be ensured.

xvii) Effectiveness of ICDS in preventing and reducing under-nutrition and promoting young child survival and development roles of ICDS & Health functionaries-ANMs, AWWs, ASHAs, AWHs and their team work with greater community involvement, especially at critical contact points will be reviewed and strengthened.

xviii) Nutritious and safe food through Public Distribution System (PDS) for households especially the unreached women and children with high nutritional vulnerabilities as envisaged in National Food Security Act, 2013 will be made available. Management of institutions of food/grain banks by women Groups (SHGs) could be considered to enable uninterrupted supply of food grains even in times of natural/man-made calamities or disasters such as flood, drought etc.

II. Education

i) Pre-School education at the Anganwadi Centres will be strengthened and efforts will be made to improve access to pre-primary education for girl children by involving the community and sensitizing the parents. This will improve children's communication and cognitive skills as a preparation for entry into primary school. This will help older children, particularly girls, to attend schools and prevent their poor performance and early dropouts by making them free from the responsibility of sibling care.

ii) In conformity with the Right to Education (RTE) Act, 2009, ensuring implementation of quality elementary education across board for all children including differently abled children and other marginalised children will remain a priority. Every effort will be made to effectively implement the RTE Act, 2009 by using the education cess particularly in addressing the infrastructure gap, availability of adequately trained teachers, promoting safe and inclusive school environment etc. in remote and tribal areas.

iii) Priority will be accorded to increased enrolment and retention of adolescent girls in schools, at post primary level it will be done through provision of gender friendly facilities like functional girls' toilets, and higher recruitment of women teachers. Promotion of skill development, vocational training and life skills as a part of the secondary school education curriculum for adolescent girls and young women will be given importance. Efforts will be made to address the repetition rates for girls including those from the marginalised communities with varying underlying causes to arrest drop-out. A mission mode approach for literacy amongst women is envisaged.

iv) Efforts will be made to provide a supportive environment in schools and colleges through a responsive complaint mechanism to address discriminatory attitudes within the organisation and in practice, particularly on the issue of sexual harassment and intimidation of girls and young women. Opportunities for recreation and participation in cultural activities will be promoted.

v) Continued efforts will be made for the gender sensitization of the faculty and curriculum, content and pedagogies for an understanding of concepts of masculinity and femininity and gender stereotypes. Gender champions in schools and colleges will be promoted to ensure gender sensitivity in the educational system.

vi) There is a need for quality management of government schools in terms of teaching, facilities and standards. Monitoring and evaluation by the community, Mothers groups and SHGs will be encouraged.

vii) Children of migrant families tend to get left out of the school system and existing schemes are not effectively coordinated or implemented. Innovative and accessible educational systems will be developed, especially in large construction sites, salt pan areas, plantations, and other manufacturing zones, which predominantly employ women labour.

viii) Major constraints that prevent women from accessing higher and technical/scientific education should be identified and girls will be encouraged to take up new subject choices linked to career opportunities. An inter-sectoral plan of action will be formulated for encouraging the enrolment of women in professional/scientific courses, by provision of financial assistance, coaching, hostels, child care etc.

ix) Distance from schools, especially secondary schools is an important factor that impacts enrolment and retention of girls in schools particularly in rural and remote areas and consequently impedes girl's access to education. Innovative transportation models will be developed such as cluster pooling of mini buses, differently abled transport, tempos, autos in addition to increasing public transport frequencies.

x) Through innovative partnerships with leading universities at the international level, opportunities for higher education will be may be expanded for ensuring access and quality to girl students and staff and also for nurturing of talent and entrepreneurship to contribute to the national development challenges.

xi) Distance education plays an important role in providing opportunities of higher education to women of all ages. Universities and academic institutions will be encouraged to launch online distance education courses to promote skill development and entrepreneurship for all women including those who had a break in their educational attainment.

xii) Adult literacy will have an added objective to link literacy programs to life skills, financial literacy, education on rights, laws, schemes etc. in partnership with government schemes such as NRLM

xiii) Efforts will be made to remove the disparities in access to and proficiency in information and communication technology (ICT), particularly between socio-economically advantaged and disadvantaged children, and between rural and urban schools as the use of ICT has now become pivotal for the entire education system. Public-private partnerships (PPP) will be adopted for building ICT infrastructure, developing applications and locally relevant content using gender-sensitive language, operations and maintenance and developing the capacity of teachers required for harnessing the full capacity of ICT productive tools.

xiv) Regular Audit on a continuous basis, of the various schemes and incentives, subsidies that are being offered to promote girls education will be undertaken to assess whether these interventions have resulted in transformative changes.

III. Economy
- Poverty Women constitute majority of population affected by poverty. Efforts for assessment of the incidence of poverty by gender estimates will be done as household estimates do not provide gender poverty estimates. Relation between gender and poverty dynamics will be addressed. Since poverty head count ratio is not sex disaggregated, alternative gendered pilot surveys to address intra-household differentials in wellbeing will be undertaken. All poverty eradication programmes will give focus to women participation.

Raising visibility
i) Increasing the participation of women in the workforce, the quality of work allotted to them and their contribution to the GDP are indicators of the extent of women being mainstreamed into the economy. Important

macro-economic policies will be engendered and mainstreamed so that women's concerns are adequately reflected and they benefit equally from the fruits of development.

ii)   Gender wage gap across rural and urban, agricultural and non-agricultural jobs, regular and casual employment will be addressed. Ensuring pay parity, satisfactory conditions of work are critical subjects for insecurity for women particularly in the informal employment. This growing informalisation and casualization of women's work / labour will also be adequately addressed.

iii)   Fiscal and monetary policies will be analysed from gender perspective since they have impending impact on the lives of women. The gender affirmative role of direct taxation will be further enhanced through various incentives like reduction in stamp duties for women if assets are registered in their name, lowering of income tax slabs for women etc.

iv)   Financial inclusion of women needs to be universalised so that women gain a financial identity, have access to financial services such as credit sources, saving services, insurance, pension schemes aimed towards poor women (with contributions), special financial literacy programmes for the poor women, and also availing of the transfer benefits and subsidies that are offered by the government. All financial inclusion schemes will incorporate monitoring and evaluation mechanisms to assess gender outcomes to women and in particular to the women belonging to the marginalized and vulnerable sections.

v)   Recognising that trade agreements are not gender neutral and that differential impact of trade policies on gender exists, especially for women working in agriculture, food processing, textiles etc., A full review of all existing trade treaties and agreements from a gender equity dimension will be made. Future negotiations should be backed by Gender Trade Impact Assessment of policy and agreements on women's wages, health and livelihood.

vi)   Women undertake the bulk of unpaid care work such as looking after and educating children, looking after older family members, caring for the sick, preparing food, cleaning, and collecting water and fuel etc. This unequal burden of unpaid care undermines women's participation in economy. Recognizing women's unpaid work in terms of economic and societal value, household surveys will be undertaken to assess the gender inequality in the household work and undertake suitable strategies to integrate unpaid work with the major programmes. Further measures will be undertaken to free woman's time for paid work through time-saving technologies, infrastructure, child/parental care services (Crèches) and child care/parental leave.

vii)   New Challenges such as increased inter-state migration, changing labour markets, meeting aspirations of the growing literate women workforce, and rapidly changing technology for women in labour force will be addressed through adequate/new skill development programmes.

viii) Identification of differently abled women through support of family, community, schools etc. and other stakeholders will be promoted along with family counselling and education to enable them to assist their differently abled members. Care-giver support programs for people with disabilities will be planned and made available for example, at the community, panchayat or municipal level. Special provisions under various rural and urban livelihood schemes for women with different forms of disabilities will be made.

ix) In order to prevent marginalization of women migrant workers at their place of destination, a system of new registration or portability of entitlements such as ration cards and identity papers from source place will be ensured particularly under PDS. Registration of tribal migrants by Panchayats will be ensured. Special efforts will be made to safeguard the interests of migrant tribal worker especially domestic workers by registration of migrant domestic workers under the Unorganized Sector Social Security Act 2008. The system of monitoring and accountability of placement agencies for domestic workers will be strengthened.

Δ Agriculture

i) Gender equity is an important concern for sustainable agricultural development. With increasing feminization of agriculture, women will be recognized as farmers in the agriculture and allied sectors and related value chain development. Efforts will be made to support women farmers in their livelihoods, their visibility and identity, secure their rights over resources, ensure entitlements over agricultural services, provide social protection cover.

ii) Concerted efforts will be made to ensure that the schemes/programmes for training women in soil conservation, social forestry, dairy development, horticulture, organic farming, livestock including small animal husbandry, poultry, fisheries etc. are expanded to benefit women working in the agriculture sector. Attention will be given to ensure the availability of extension services offered by different line departments to women farmers. Efforts will be made to utilize skills and capacities of successful women farmers as last mile extension workers and trainers or 'Krishi Sakhis' in order to extend agriculture extension services.

iii) Women have been traditionally known to conserve genetic diversity (seed banks, selection and preservation) and champion good agricultural practices. Women collectives like SHGs, cooperatives will be encouraged and incentivized in following sustainable agriculture practices. Agriculture for Nutrition campaign will be launched in selected districts where nutrition indicators are poor in popularizing cultivation of nutritional crops, horticulture products and traditional varieties. Procurement of such crops will be prioritized here, so that the same can be used in Anganwadis, for supplementary nutrition and in school mid day meals. Skill development for forest-based, livestock-based, poultry and fisheries-based livelihoods will be encouraged as part of the inclusive strategy in agriculture.

iv)   Legal provisions of the relevant Acts will be effectively implemented to ensure the rights of women to immovable property. Exploitation that arises out of land ownership by women such as witch hunting will also be addressed.

v)   Regarding resource rights of women, efforts will be made to prioritize women in all government land redistribution, land purchase and land lease schemes to enable women to own and control land through issue of individual or joint land pattas. In the case of private land, joint registration of land with spouses or registration solely in the name of women will be encouraged along with measures such as concessions in registration fee and stamp duty etc. to incentivize land transfers to women.

vi)   Women farmers' collective farming enterprises will be incentivized, by providing support for post-harvest storage, processing and marketing facilities. Where this is done by leasing in land, appropriate changes in tenancy laws will be facilitated. Institutional and funding support for the formation of women producers associations and existing women's federations/cooperatives to process, store, transport and market farm produce, milk, fish, crops etc. will be provided.

vii)   Efforts will also be directed to accurately capture and reflect women's work in the agriculture and allied sectors including gender differentials in wages, gender sensitive social security policies at regular intervals to inform Research and Development, policies and programmes.

viii) Gender-disaggregated land ownership database at all levels starting from the revenue village upwards will be collected and maintained for more focused interventions.

ix)   Steps will be taken to involve women farmers in on-farm participatory research for agricultural technology and development of women friendly implements/ tools. Database of women friendly technologies/ equipments available for all stages in the agriculture value chain for bulk purchase with list of manufacturers will be developed.

x)   Wives of farmers who committed suicide on account of failure of crops or heavy indebtedness are highly vulnerable and are left behind to take care of their children and family. Special package for these women that contains comprehensive inputs of programs of various departments/ Ministries like agriculture, rural development, KVIC, MWCD will be provided for alternative livelihood options.

Δ Industry, Labour and Employment (Skill Development, Entrepreneurship)

i) As the Indian economy grows and more new and innovative initiatives take place in the public and private domain, women have to have a fair share of these development gains. Indicators of mainstreaming women in the economy such as participation of women in workforce, type of work allotted to them and their contribution to GDP will be developed and monitored.

ii) Suitable strategies will be developed and implemented to ensure that women have equal opportunities to enter and enjoy decent work, in just and favourable environment, including fair and equal wages, social security measures, occupational safety and health measures. Appropriate steps will be taken to facilitate women workers and economic units move from the informal economy to the formal economy.

iii) Effort will be made for training and skill upgradation of women in traditional, new and emerging areas to promote women employment in both organized /unorganized sectors as envisaged in the new National Policy for Skill Development and Entrepreneurship 2015. Special Emphasis will be given to skill development of marginalised women and those in difficult circumstances in the unorganised sector, and by linking them to urban and rural livelihood programs. Special provisions will also be made for promoting re-entry of highly /technically skilled women in the job market especially for those who resign or take a break to manage the care economy.

iv) Entrepreneurial development must ensure participation of women through accelerated involvement in various sectors through programmes and schemes of various departments/Ministries while identifying their needs such as access to credit, technology, market etc.

v) Specific efforts will be made to increase work participation of women in the organised and industry sectors. The availability/creation of part-time jobs and arrangement of flexi-hours in the organized sectors will be promoted. Provision of affordable housing and gender friendly facilities at workplace will be made as more women tend to migrate to cities and metros for work.

vi) A review of Labour Acts and policies for increasing female work participation and for eliminating discrimination and promoting equity will be undertaken. Suitable policies will be introduced to promote workforce participation in terms of parental leave and child and elder care.

vii) Effective safety nets mechanisms will be formulated for migrant women such as those working in construction, domestic servants; brick kilns plantations, along with protection of their entitlement of benefits such as BPL and ration cards.

viii) A mechanism will be put in place for monitoring the compliance of mandatory laws like Maternity Benefit Act and The Sexual Harassment of Women at Work Place (Prevention, Prohibition and Redressal) Act, and display of the rights and benefits of female employees provided by the organization. Provisions such as natal and post-natal benefits, child care facilities, flexitime, housing which impact women's productivity will also be encouraged.

Service Sector

i) Women's participation in the upcoming services such as information based industries, telecommunication, infrastructure, customized highly skilled business services, software-designs; computer programming and financial services (Banks and insurance) will be encouraged. Skills and work incentives for frontline workers which rely heavily on female labour in health and education will be strengthened.

ii) The service sector will encourage equal employment opportunity through jobs/enterprises for women especially in high paid jobs to post graduates and professionally qualified women.

iii) Conducive infrastructural facilities such as toilets, restrooms, child care facilities at workplaces will be ensured for providing a safe and encouraging working environment for women. Provisions for women friendly infrastructure should be part of Urban /Panchayat planning processes. Efforts will be made to improve accessibility of safe public transport for women.

iv) Enable women to access the formal banking system with their own collateral and develop a strategy for the financial institutions to address the significant gender gap by removing the constraints for accessing private finance.

Δ Science and Technology

i) Technological needs of women, in both urban and rural areas as well as across various sectors will be addressed. Use of technology as a tool to increase employment, reduce drudgery, improve access to health, education, and communication services and political participation etc. will be compiled and suitably incorporated in training and best practices manuals, and widely disseminated in all training programs.

ii) Since women greatly benefit from ICTs, mobile telephone applications will be proactively used as a tool for mass communication and dissemination of information on legal rights, payments under wage employment schemes, subsidies, pension payments, markets etc. Efforts will be made to collect gender based data through mobile phones to feed into policy prescriptions.

iii) Enabling mechanisms will be institutionalized to encourage girl students/women to enter into the areas of science, information and communication technology, for ensuring technical training, its access and usage through e-education in rural areas and to serve as a means for income generation.

vi) To enable women SHGs, cooperatives, federations, CBOs, NGOs to take active part in technology dissemination, suitable training manuals will be prepared and cascading training programs organised at multiple levels.

IV. Governance and Decision Making

i) Establish mechanisms to promote women's presence in all the three branches of the government including the legislature, executive and judiciary. Women's participation in the political arena will be ensured at all levels of local governments, state legislations and national parliament with at least 50% reservation for women in local bodies and 33% in state assemblies and parliament to provide more responsive, equitable and participatory development.

ii) Increase the participation of women in civil services, judiciary and in corporate boardrooms through appropriate modules for guidance and counselling, coaching, provision of financial incentives and quotas.

iii) Increase the participation of women at all levels such as in trade unions, political parties, interest groups, professional associations, and businesses/private sector.

iv) In order to achieve women's full participation and representation at all levels, maintain gender disaggregated data to track and assess progress, or serious inconsistencies.

v) Strengthen the Administrative Training Institutes (ATIs) to systematically train the civil servants on gender issues to efficiently and effectively respond to the gender based challenges created by the rapid economic growth, devolution of funds, enhanced transparency through the right to information, globalization, climate change and extremism and so on.

vi) To enable women SHGs, cooperatives, federations, CBOs, NGOs to take active part in decision making, and promoting women's rights, capacity building exercises and training programs will be undertaken.

vii) Quality of women's representation will be improved through greater capacity building on aspects of decision making and women's rights and legislations.

V. Violence against Women

i) Efforts to address all forms of violence against women will be continued with a holistic perspective through a life cycle approach in a continuum from the foetus to the elderly starting from sex selective termination of pregnancy, denial of education, child marriage to violence faced by women in private sphere of home, public spaces and at workplace. It will identify and combat violence and abuse through a combination of laws, programs, and services with the support of diverse stakeholders.

ii)   Efforts to improve Child Sex Ratio (CSR) will be continued through a judicious combination of effective implementation of the Pre-Conception and Pre-Natal Diagnostic Techniques (PCPNDT) Act, 1994, and advocacy through awareness and sensitisation to change the mindsets by involving communities and the stakeholders for valuing the girl Child. Special measures to combat violence and crimes against adolescent girls in public and domestic spaces will be adopted.

iii)   Trafficking of women and children is a cause for concern and will receive prioritized attention. Requisite steps for prevention of trafficking at source, transit and destination areas for effective monitoring of the networks of trafficking will be given a priority. Existing legislations/schemes for prevention, rehabilitation of victims of trafficking will be suitably strengthened. Efforts will be made to raise awareness regarding the subtle and violent nature of sex trafficking and how women and children subjected to this crime are victimized through coercion.

iv)   There is need for effective implementation of The Persons with Disabilities (Equal Opportunities, Protection of Rights and Full Participation) Act, 1995 to ensure that all provisions of the Act are benefiting differently abled girls and women. To prevent violence and sexual exploitation of the differently abled, focussed advocacy and sensitisation of various stakeholders such as law enforcement, judiciary, panchayats will be undertaken.

v)   Strict monitoring of response of enforcement agencies to violence against women will be put in place. Efforts will be made to ensure speedy /time bound trial of heinous crimes against women. Alternate dispute redressal systems such as family courts, Nari Adalats etc., will be strengthened.

vi)   Efforts will be made to increase the representation of women in judicial positions across the board.

vii)   Effective mechanisms for network and convergence of relevant institutions/agencies like National Legal Services Authority (NLSA), District Legal Service Agency (DLSA), National Commission for Women and Ministry of Women and Child development will be strengthened for providing easy and affordable access of justice to woman. NLSA, DLSA will create linkages with supportive institutions such as Shelter homes, One Stop Centres in order to give required legal aid to women staying in these homes.

viii)   Efforts will be made to streamline data systems through review of various data sources (Census, NFHS, NSS, NCRB) to develop a compatible and comprehensive data base on Violence Against Women.

ix)   Engaging men and boys through advocacy, awareness generation programmes and community programmes will be undertaken.

x)   Gender specific training incorporating gender sensitivity and a thorough briefing on the specific laws for women will be undertaken continuously for all ranks and categories of police personnel. Training for the judiciary, judicial schools, and all legal practitioners, will be accorded a priority for developing the specialized skills needed to investigate and prosecute cases related to gender based violence.

VI. Enabling Environment
• Housing and Shelter
i) Gender perspective in housing policies, planning of housing colonies and in the shelters both in rural and urban areas will be given a priority. Special attention will be given for providing safe, adequate and affordable housing and accommodation to women in urban areas including single women, homeless, migrants, women heads of households, working women, students, apprentices and trainees etc.

ii) State run temporary or permanent shelter homes will have improved living conditions for women survivors of domestic violence or any other forms of violence to feel safe and secure. Efforts will be also be made to make slums and informal settlements safe for women.

iii) Central or State run housing scheme that provides for a house in the name of women will be universalized.

• Drinking Water and Sanitation
iv) Ensuring safe drinking water and sanitation will be considered critical for the health of women. In this context, accessible and operational toilets for women within the house or at the community level will be universalised. Women's groups may be involved for the operation and maintenance of women community toilets in rural areas. Availability and access to adequate water and sanitation facilities including menstrual hygiene support for adolescent girls will be facilitated to improve retention of girls in the school.

v) Efforts to educate women and girls about the dangers of unhygienic practices will be continued. Sustained efforts will also be made to increase the running water facilities in schools to improve menstrual hygiene among adolescent girls and their retention in schools. Sewage disposal facilities in schools/households/community places especially in rural areas and urban slums will be encouraged.

vi) To mitigate women's water burden, future programmes/projects should be designed keeping in view the women as water users. Water conservation programmes need to be initiated to generate awareness on conservation methods such as rain water harvesting by involving women groups. Women should be trained as water manager for better utilization of water.

vii) Water resource management strategies should include a gender perspective which inter-alia values and reinforce the important role played by women in acquiring, conserving and using water. Women will also be included in the decision making process related to waste disposal, improving water and sanitation systems and agricultural industrial and other land use projects that affect quality and quantity of water. Women's participation will be ensured in the planning, delivery and maintenance of such services.

- ## Media
  i)   Gender parity in the mass media i.e. print and electronic media, advertising world, film sector and new media will be promoted by increasing the presence of women in the decision making positions.
  ii) Encourage the entry of women in media industry through promotion of journalism and mass media courses and ensuring adherence to equitable work conditions. Setting up women media centres to provide technical training and skill building will be encouraged.
  iii) Gender Sensitisation and non-discrimination in portrayal of women in all forms of media and use of gender sensitive language will be advocated to ensure that women are not represented in a demeaning or stereotypical manner. Private sector media networks will be sensitised and encouraged to portray empowering images of women with self-regulatory mechanisms.
  iv) Encourage the use of new media applications of social media like internet, facebook, twitter and mobile phone communication and information systems to benefit women while at the same time provide enough safeguards to protect the dignity and safety of women.

- ## Sports
  i)   Social practices and physical differences between men and women usually make it imperative that separate but equal facilities be made available to girls and women in terms of sports equipment, scientific support, medical support, diet/nutritional support, financial support and competition exposure. Financial support in terms of sponsorship/ coaching/ equipment for budding talent of girls especially in rural areas will be promoted.
  ii) Conscious efforts will be made to ensure gender parity in the induction of trainees, trainers, recruitment of coaches under various schemes of Sports Authority of India (SAI). Sports competitions that start from block level onwards to district, state and national level will promote participation of women. Effort to make existing institutional sports facilities available to sports women on a neighbourhood basis will be strengthened.

- ## Social Security
  i)   Efforts will be made to strengthen the existing supportive social infrastructure for women especially the vulnerable, marginalized, migrant and single women. This is imperative in view of their increasing life span, more conservative investment habits, variability

in labour force participation and the role in home based production or unpaid work. Attention will also be paid to the needs of domestic workers and appropriate steps taken to ensure overtime pay, annual paid leave, minimum wages and safe working conditions.

ii) Concerted efforts will be made towards strengthening social security and support services like insurance products, pensions, travel concessions, subsidies, benefits under BPL systems, childcare, crèches, working women hostels, shelter homes. Corporate sector may be involved on a long term basis in the financing of the insurance component under the Corporate Social Responsibility (CSR).

iii) Among the various aspects of social security, measures will be taken to ensure effective implementation of provisions of maternity benefits such as leave and nursing breaks under the Maternity Benefit Act (1961).

iv) In the context of social security and poverty alleviation the Public Distribution System should be mandatorily linked to the data generated by the Socio-Economic Caste Census for ensuring more inclusive and transparent distribution of food to the disadvantaged women in remote rural and tribal areas.

- **Infrastructure**
  i) Efforts will be made to provide affordable and improved conventional transport services on feeder roads and the potential for women's group/community based low-cost transport schemes will be explored. More women transport professionals will be trained and promoted to safeguard women's safety and security.

  ii) To promote safety of women, all urban planning and smart city projects will mandatorily include safe and gender sensitive infrastructure and facilities. Urban Safety Audits will be conducted periodically to ensure all gender safe measures are provided.

  iii) Similarly, in rural areas, panchayat functionaries' orientation will include capacity building on development of women friendly and safe infrastructure. They will be persuaded to allocate at least 10 percent of their budget towards the development of women friendly and child friendly infrastructure.

VII. Environment and Climate Change

i) As women are highly affected by climate change, environmental degradation, distress migration and displacement in times of natural calamities, policies and programmes for environment, conservation and restoration will compulsorily incorporate gender concerns. An integral part of this discourse will be to enable equitable ownership control and use of natural resources and secure the asset base of marginalised poor women to counter poverty and climate shocks.

ii) Mitigation and management of natural disasters is very important from the gender perspective as floods, droughts etc. impact women severely exposing them to social and economic vulnerabilities. Holistic gender specific strategies in disaster management and prevention will be formulated and implemented aiming at their safety, well-being, health and rehabilitation.

iii) Measures will be taken to comprehensively address the adverse implications of crop selection, use of chemicals and pesticides etc on women's health and their surrounding environment.

iv) Environmental friendly, renewable, non–conventional energy, green energy sources will be promoted and made affordable and accessible to rural households for their basic household activities.

v) Women participation will be ensured in the efficient use and spreading the use of solar energy, biogas, smokeless chulas and other technological applications to have positive influence on their life styles and a long term impact on meeting sustainable development goals. Micro-enterprises based on environment friendly technologies, organically grown produce will be promoted to provide viable livelihoods options to women.

vi) All aspects of energy planning and policy-making must include gender dimensions and actively advance women's leadership. Specific attention would be paid to remove the barriers to women executives, entrepreneurs and employees within the energy sector as sustainable energy future requires a diversity of voices, and the solution has to be encompassing not only to break the cycle of poverty but to respond to the increasing workforce demand of the expanding energy sector.

vii) Though gender roles in using forest resources differ widely depending upon the region as well as socioeconomic class and tribal affiliation, rural women's association with the forests varies from gathering and processing of Non-timber Forest Products (NTFP) to wage employment, production in farm forestry and management of afforested areas in the community plantation. Women sustain household economies, procuring food, fodder, Minor Forests Produce (MFP), fuel-wood and other needs such as medicinal plants from forests. Efforts will be made to recognize their rights under the Forest Rights Act as individuals and communities and their roles in forest governance will be strengthened.

viii) Sustained efforts will be made to increase the capacity of local women, SHGs, elected women representatives to mitigate and adapt to climate change while increasing their knowledge on the subject.

Δ Emerging issues

i) Redistribution of gender roles is imperative in a scenario where women are expected and aspire to contribute to the economic development of the country. Efforts will be made to prepare family-friendly policies, which provide for childcare, dependent care and paid leave for women and men both in organized and unorganized sectors to help employees balance work and family roles.

ii) Given the plurality of the personal laws a review is required of the personal and customary laws in accordance with the Constitutional provisions. This will enable equitable and inclusive and just entitlements for women.

iii) With technological advancement, there has been incidence of frauds, misuse of information uploaded on the cyberspace and hence there is a need for developing protective measures for citizens keeping in view that victims of such frauds are largely women.

iv) As more women are taking the recourse of artificial reproductive techniques, efforts will be made to ensure the rights of these women adopting these techniques i.e. surrogates mothers, commissioning mother along with children born as a result will be protected.

v) To recognize special needs of single women including widows separated, divorced, never-married and deserted women. A comprehensive social protection mechanism will be designed to address their vulnerabilities, create opportunities and improve their overall conditions.

vi) To create ecosystem for women to participate in entrepreneurial activities, take up decision-making roles and leadership in all sectors ofthe economy.

Δ The promotional schemes available in the country in order to develop women entrepreneurship are as follows.

* Mahila Nidhi.
* Mahila vikas Nidhi
* Priyadarshini yojana .
* Trade-related entrepreneurship assistance and development (TREAD).
* Special programs conducted by the SIDO (small industries development organization)
* CWEI (the consortium of women entrepreneurs of India.
* WIT (women India trust).
* SWEA (self-employed women association) .
* SHG's (self-help group)
* FTWE (federation of women entrepreneurs)
* Income generating schemes by Department of women and child development.
* KVIC (khadi villages industries commission)
* DIC (District industrial centers)
* Women cell
* Women industries fund schemes

Δ Empowerment can be at two levels:

Personal: a sense of the self or self-efficacy wherein individuals have the confidence and capacity to look at internalized oppression. In addition, personal empowerment gives individuals the ability to negotiate unequal relationships and influence decision-making.

Collective: where individuals work together, for example, through self-help groups and co-operatives, or at a larger scale, through networks, alliances and social movements, to achieve a more extensive impact than each would have done alone. Group efforts involve collective awareness and capacity building by providing spaces for women (and men) to come together and understand their (disempowered) situation.

Δ Creating Empowerment

- The prime target of empowerment must be adult women and, in the context of social justice and transformation, they must be low-income adult women. Within this group, authoritarian behaviors by husbands in the home make families and households in general a terrain that serves the maintenance rather than the transformation of unequal gender relations.

- A prerequisite to empowerment, therefore, necessitates stepping outside the home and participating in some form of collective undertaking that can be successful, thus developing a sense of independence and competence among the women. The creation of a small, cohesive group, with which its members may identify closely is paramount. We know that because of the small scale and voluntary nature of these associations many members gain valuable experience and confidence in both leadership and membership tasks. The central activity of the group could vary; it could be literacy activity, income-generation, mutual basic needs support, etc. Whatever the objective, the group activity should be designed so that its process and its goal-attainment foster the development of a sense of self-esteem, competence, and autonomy.

- Empowerment will go through a series of phases. Awareness of conditions at the personal and collective levels will lead to some public action, however small. Following from this beginning there should occur a renegotiation of family conditions. As women become more available for public action, they should be able to place more demands upon the state.

- Women can attain empowerment through different points of departure: emancipator knowledge, economic leverage, political mobilisation. While many poor women work outside the home to support their families and the tasks they perform are exhausting and meagrely rewarded, access to income improves their authority in the home. Working women, regardless of how inferior their position and small their income, have a greater sense of control over their lives and more power and control over resources within the family than nonworking women (for a detailed ethnographic study comparing working and nonworking women in six communities in the Dominican Republic, see Finlay, 1989). A study of 140 women home workers in Mexico City by Beneria and Roldan (1987) found that while no simple relationship existed between women's economic resources and decision making, paid work increased the women's self-esteem and wives who made a considerable contribution to household expenditures (more than 40 percent) had augmented their domestic and conjugal decision making.

- Mothers' clubs make possible the creation of free and socially accepted spaces for women. Although many of their activities do not seek transformational objectives, the clubs can provide fertile ground for empowering processes. The crucial point about these mothers' clubs, usually created under religious auspices in Latin America and Africa, is that they represent a large number of the collective spaces already occupied by women.

- Literacy skills can also be empowering, but they must be accompanied by a process that is participatory and content that questions established gender relations, features that, unfortunately, do not characterise the great majority of literacy programs. Nonetheless, evidence from Asia and Latin America indicates that women with newly acquired literacy skills have moved into self-help organizations ranging from neighborhood soup kitchens to public health groups (Bown, 1990; Stromquist, 1993).

Δ The Pedagogical Rationale for Empowerment

The creation of critical minds requires a physical and reflective space where new ideas may be entertained and argued, and were transformational demands may occur outside the surveillance of those who may seek to control these changes.

Sara Evans, an experienced member of the feminist movement in the U.S., reviewing the social roots of feminism in the 1950s and 1960s, concluded that prerequisites to an "insurgent collective identify" are the following:

1. Social spaces within which members of an oppressed group can develop an independent sense of worth in contrast to their received definitions as second class or inferior citizens;

2. Role models of people breaking out of patterns of passivity;

3. An ideology that can explain the sources of oppression, justify revolt, and provide a vision of a qualitatively different future;

4. A threat to the new-found sense of self that forces a confrontation with the inherited cultural definitions-in other words, it becomes impossible for the individual to "make it on her own" and escape the boundaries of the oppressed group; and finally

5. A communication or friendship network through which a new interpretation can spread, activating the insurgent consciousness into a social movement.

- The need for a social space, a "free institutional space," for people with a shared condition was first discovered by people in the U.S. movement of the political left of the 1970s. Interestingly, recent findings from organizational behaviour support this strategy. Organizational theory and empirical evidence support the notion that knowledge is socially constructed. A process of mobilization and collective action develops a shared cognitive system and shared memories. These forms of organizational cognition, which call for the understanding of events, open the opportunity for social interpretation as well as the development of relatively dense interpersonal networks for sharing and evaluating the information, thus creating effective learning systems:

- Organizational learning can be relatively low level or single loop, involving only minor adjustments and fine tuning of existing organisational images and maps. Conversely, it can be reflected in the alteration of existing norms, assumptions, and values that govern action. Such learning is referred to as high-level or double-loop learning.

- This collective learning, which draws upon the theory of social learning of Albert Bandura, has been argued to be one of the greatest benefits from participatory evaluations in education (Cousins and Earl, 1992). In my view, the rationale of learning that occurs in women's groups is the same.

- Empowerment can succeed only if it is a mode of learning close to the women's everyday experiences and if it builds upon the intellectual, emotional, and cultural resources the participants bring to their social space. In the Chilean and the Brazilian projects mentioned earlier there was a clear focus upon knowing the experiences of the women in their everyday life; there was an equally strong focus on making those experiences collective. This discussion of everyday life has a number of consequences. When women talk to other women about their personal experiences, they validate it and construct a new reality. When women describe their own experiences, they discover their role as agents in their own world and also start establishing connections between their micro realities and macro social contexts. It should be clear that the discussion of personal lives, of needs and dreams, necessitates of a friendly, receptive social space. Here, the task of a group facilitator becomes essential because this person must create a participatory process which provides constant encouragement and support to the members. The role of the facilitator is not an easy one; training to create and maintain an empowerment process is necessary.

Schrijvers proposes four criteria to assess an existing degree of women's autonomy:

1. women's control of their own sexuality and fertility; forms of shared mothering, between women or between women and men;
2. A division of labour which allows women and men equal access to, and control over the means of production;
3. Forms of cooperation and organization of women which will enable and help them to control their own affairs; and
4. Positive gender conceptions which legitimate women's sense of dignity and self-respect, and their right to self-determination (Schrijvers. p.3).

These criteria approximate to the notion of multi-faceted empowerment. But they need to be toned down to address the concrete form in which empowerment is likely to take place, that is, within a specific project or program that is organizationally bounded. In this case, empowerment should be assessed by the number of facets the project addresses (cognitive, psychological, economic), the changes it brings in terms of women's individual understanding and collective action, the strength and stability of their organization, the renegotiation of authority it enables at the household and community levels, and the range of objectives it identifies for future action.

Δ Empowerment and Education

In talking about empowerment activities I have focused exclusively on adult women and therefore considered only non-formal education. Does it mean formal education has no empowering role for girls? Formal education has substantial contributions to make to an improved gender identity through the removal of sexual stereotypes in textbooks, the fostering of positive gender identities through the curricula, the retraining of teachers to be gender sensitive, and the provision of non-sexist guidance and counseling. These elements, in my view, are crucial antecedents of empowerment, not empowerment itself. I prefer to reserve the concept of empowerment for behaviours that tie understanding to a clear plan of action to vindicate the rights of women. If the concept of empowerment is freely applied to changes that are only cognitive or psychological, empowerment would not necessarily have to be translated into a collective dimension. And in the case of women's transformation, it is imperative that social structures be rearranged.

Δ Barriers to Empowerment

While it is clear that many benefits may derive from collective action, it must also be remembered that participation in groups with a serious purpose of vindication will demand sustained involvement.

- Poor women are busy women. Not only do they spend much time and energy responding to family needs, but they also face conditions such as rigid authoritarian spouse control, violence at home, social expectations regarding motherhood, and unsafe community environments that limit their physical mobility. Under these conditions, participation is fraught with obstacles and only a few will find it possible to become available for participation. The percentage of women that will participate under these circumstances is not well known, but judging from rates of participation in related activities, particularly literacy groups which call for prolonged involvement, this proportion may be less than five percent of the possible population. Projects working on empowerment will be small in their beginnings and take a substantial amount of time to mature and solidify. Ambitious expectations of quick and mass appeal have no basis in fact. How to make it possible for women to engage in empowering activities while they face a critical everyday survival is a real challenge.

- The increased interest in empowerment comes at a time when structural adjustment policies are being implemented in many of the developing countries. There is strong evidence that these policies have had a negative impact on women in multiple dimensions of their lives, including education (see Commonwealth Secretariat, 1989, and UNICEF, 1987). In fact, the Commonwealth Secretariat report concluded that, the types of stabilization and adjustment policies followed in the 1980s have brought to a standstill many of the practical advances which women made earlier, and have actually reversed some of the most fundamental of them like education and health.

- To break some of the barriers in empowerment, the work of three sets of actors will be needed: grassroots and feminist groups to do the outreach and work with marginalized women who need support; women in development and international institutions who can provide the funds necessary to create projects and programs with empowerment features; and women in academic circles, who will contribute theoretical analysis of how gender is created and how it can be modified in society.

Δ   Indicators of Empowerment

Understanding that empowerment is a complex issue with varying interpretations in different societal, national and cultural contexts, the participants also came out with a tentative listing of indicators.

At the level of the individual woman and her household

- Participation in crucial decision-making processes;
- Extent of sharing of domestic work by men;
- Extent to which a woman takes control of her reproductive functions and decides on family size;

- Extent to which a woman is able to decide where the income she has earned will be channelled to;
- Feeling and expression of pride and value in her work;
- Self-confidence and self-esteem;
- And ability to prevent violence.

At the community and/or organisational
- Existence of women's organisations;
- Allocation of funds to women and women's projects;
- increased number of women leaders at village, district, provincial and national levels;
- Involvement of women in the design, development and application of technology;
- Participation in community programmes, productive enterprises, politics and arts;
- Involvement of women in non-traditional tasks; and
- increased training programmes for women; and
- exercising her legal rights when necessary.

At the national level
- Awareness of her social and political rights;
- Integration of women in the general national development plan;
- Existence of women's networks and publications;
- Extent to which women are officially visible and recognized; and
- The degree to which the media take heed of women's issues.

Δ Facilitating and Constraining Factors of Empowerment

Empowerment does not take place in a vacuum. In the same way that Ms. Lazo talks about women's state of powerlessness as a result of "a combination and interaction of environmental factors," one can also discuss the conditions/ factors that can hasten or hinder empowerment. As above, the listing is a preliminary one based on the discussions.

Facilitating factors
- Existence of women's organisations;
- Availability of support systems for women;
- Availability of women-specific data and other relevant information;
- Availability of funds;
- Feminist leadership;
- Networking; favourable media coverage;

- Favourable policy climate.

Constraining factors
- Heavy work load of women;
- Isolation of women from each other;
- Illiteracy;
- Traditional views that limit women's participation;
- No funds;
- Internal strife/militarization/wars;
- Disagreements/conflicts among women's groups; ustructural adjustment policies;
- Discriminatory policy environment;
- Negative and sensational coverage of media.

Δ Strategies for the Future

Empowerment through education is ideally seen as a continuous holistic process with cognitive, psychological, economic and political dimensions in order to achieve emancipation. Given the complexity of political, societal and international interrelations, one has to systematically think about the strategies and concrete proposals for future action if one hopes to achieve such a goal. A set of strategies on education, research/ documentation, campaigns, networking, influencing policies, training and media was developed by the participants. As can be seen from the listing, the strategies are inter-related to each other.

Δ Education

The formal and non-formal education systems would need to be considered. It would be important to analyze the gender content and to ascertain the manner in which it is addressed/not addressed in the educational system. On the basis of the analysis, curriculum changes would need to be brought about. Likewise it would be important to reorient the teachers on gender issues so that overall gender sensitisation in the educational system could be brought about.

In concrete terms, this would mean:
- Reorienting and re-educating policy makers;
- securing equal access for boys and girls in education;
- holding workshops/seminars for teachers;
- revising teaching materials;
- producing materials in local languages;
- implementing special programmes for women in the field of Adult Education;

- incorporating issues such as tradition, race, ethnicity, gender sensitisation, urban and rural contexts in the programmes;
- raising awareness on the necessity for health care;
- politicising women to show them how macro level mismanagement is responsible for their loss of jobs; and
- focusing on parents as role models.

Δ Research/Documentation

The importance of doing participatory and action research was underscored. It was considered important to organize workshops to train grassroots women to conduct participatory research where they could develop skills to critically analyze their existing conditions. This will facilitate their organizing for collective action.

While participatory research was considered to be important, it was recognized that traditional quantitative research was also necessary. The guiding principle, however, was to share the results with the women in a language and manner that was understandable to them.

Research as a strategy would therefore entail:

- disseminating information;
- producing and disseminating information leaflets regarding women's rights;
- referring to women in all national and UN statistics;
- collecting oral history of women;
- documenting and analyzing successful and failed programmes of the women's movements;
- analyzing successful advocacy cases in order to learn about the arguments that persuade policy makers;
- collecting cross-cultural case studies;
- Constantly evaluating research; and
- involving women as agents (instead of objects) of research.

c. Campaigns

If one is to have an effect in society, it is important to undertake campaign and lobby activities that will put the issue of gender in the minds of the legislators, policy-makers and the larger public. This will therefore

- pushing for a dialogue between stake holders;
- raising gender issues within the national policy arena;
- pressuring to upgrade women's bureaus (which are a result of the UN Decade for Women) into ministries of women's affairs;
- lobbying for sex-equity and affirmative action legislation;

- lobbying for "counter structural adjustment policies;" organizing pressure groups (like "Greenpeace");
- using consumer power for boycotts; securing access to information;
- demanding child care centres; and
- producing videos and CDs, T-Shirts etc.

d. Networking

Through networking, it would be possible to share experiences and learn from one another. In this manner, understanding and solidarity among women's organisations, development organizations (governmental/non-governmental) and multilateral agencies could be forged. This would therefore entail networking at the national, regional and international levels. Moreover, at the international level, South-South linkages were considered to be particularly important.

- organizing at least one meeting a year of gender sensitive organizations;
- bringing together donor agencies, governments and NGOs;
- setting up a north-south dialogue and collaboration;
- setting up a south-south cooperation and exchange;
- linking women's movements all over the world;
- establishing alternative credit schemes that offer women access to funds.

e. Training

In our societies, there is a gender division of labour which dictates the kind of training one acquires. If one talks about women's empowerment, it is important that women have access to the different training opportunities previously denied them. This therefore means:

- preparing for jobs that are usually not open to them;
- providing income-generating projects that are market-oriented (not welfare-oriented projects); and
- training capable female leaders at all levels.

f. Media

Considering the attitudinal barriers in traditional societies and the role which the mass media play in reinforcing them, the following strategies were advanced:

- organizing mass media campaigns to raise awareness;
- creating a social climate friendly to women's issues;
- resisting the tendency to send women back to the kitchen; and
- disseminating information about conferences that will take place in the coming years.

Δ Strategies for rural women's empowerment (TYPES)

Since empowerment is a multi-faceted concept, strategies for facilitating women's personal and collective empowerment are diverse and include:

- Legal empowerment: through legislative action and its enforcement-reform in personal laws, policy on sexual harassment at the workplace (Bill being drafted), women's empowerment policies (Centre and some state governments) and laws to protect women against violence.

- Political empowerment: Facilitating women's participation in governance at different levels, for example, through one third reservation in local government under the 73ro and 74th constitutional Amendments.

- Economic empowerment: Enhancing women's access to and control over productive resources/endowments and benefits (both private and collective),for example, through the formation of SHGs, micro-credit activities, the promotion of entrepreneurship skills, access to the market.

- Social empowerment: Increasing women's ability to gain control over their bodies, fertility, sexuality and their identities through access to health (reproductive rights) and education (literacy and life-skills).

Essentially all these approaches towards empowerment are interconnected and emphasize the need for policy advocacy, legislative change, capacity building and strategies which facilitate women's participation and collective mobilisation at different levels while critically questioning existing gender relations. The Millennium Development Goals recognise that women's empowerment is central to poverty alleviation and development objectives. Goal 3 calls for the promotion of gender equality and empowerment of women through the elimination of gender disparities in primary and secondary education by the year 2015. However, transformatory change calls for strong leadership and gender-progressive organisational change that involves both men and women in a process which by its very definition is nonlinear, complex and context-specific.

Δ Measuring empowerment

The claims for women's empowerment as the ultimate objective of many development policies and programmes leads to a demand for indicators both to reveal the extent to which women are already empowered and to evaluate if such policies or programmes have been effective in achieving their stated aims. Although a variety of indicators have been developed, given the multi-dimensional nature of empowerment it is difficult to quantify or 'Measure'. With respect to broad or societal indicators, the UN Human Development Report prepared for the women's conference in Beijing (UNDP 1995) came out with two composite indexes, namely the Gender Development Index (GDI) and the Gender Empowerment Measure (GEM). Empowerment is recognised as one of the four essential components of the human development paradigm, the others being productivity, equity and sustainability.

The GDI attempts to measure countries' achievements in the basic capabilities covered by the HDI (life expectancy, educational attainment and income), taking note of the inequalities in achievement between women and men, and penalizing for inequality. Countries with greater gender disparities will have lower GDIs compared to their HDI.

Δ Millennium Development Goals and Targets

1. Eradicate extreme poverty and hunger
    •   Reduce by half the proportion of people living on less than a dollar a day
    •   Reduce by half the proportion of people who suffer from hunger

2. Achieve universal primary education
    •   Ensure that all boys and girls complete a full course of primary schooling
3. Promote gender equality and empower women
    •   Eliminate gender disparity in primary and .secondary education preferably by 2005, and at all levels by 2015

In detail:

Goal: promote gender equality and empower women.

Target: eliminate gender disparity in primary and secondary education preferably by 2005 and to all levels of education no later than 2015.

Indicators:

•   ratio of girls to boys in primary, secondary and tertiary education
•   ratio of literate females to males of 15-24 year olds
•   share of women in wage employment in the non-agricultural sector
•   Proportion of seats held by women in national parliament.

Research has shown that education for girls is the single most effective way of reducing poverty. In this context, the elimination of gender disparity in education has been selected as the key target to demonstrate progress towards gender equality/women's empowerment. However, education alone is not enough. Progress towards gender equality in education is dependent on success in tackling inequalities in wider aspects of economic, political, social and cultural life, and this is reflected in the indicators of progress.

4. Reduce child mortality
   - Reduce by two-thirds the mortality rate among children under five

5. Improve maternal health
   - Reduce by three-quarters the maternal mortality ratio

6. Combat HIV/AIDS, malaria and other diseases
   - Halt and begin to reverse the spread of HIV/AIDS
   - Halt and begin to reverse the incidence of malaria and other major diseases

7. Ensure environmental sustainability
   - Integrate the principles of sustainable development into country policies and programmes; reverse loss of environmental resources
   - Reduce by half the proportion of people without sustainable access to safe drinking water
   - Achieve significant improvement in lives of at least 100 million slum dwellers by 2020

8. Develop a global partnership for development
   - Develop further an open trading and financial system that is rule-based, predictable and non-discriminatory. Includes a commitment to good governance, development and poverty reduction-nationally and internationally.
   - Address the least developed countries' special needs. This includes tariff and quota-free access for their exports; enhanced debt relief for heavily indebted poor countries; cancellation of official bilateral debt; and more generous official development assistance for countries committed to poverty Reduction
   - Address the special needs of landlocked and small island developing States
   - Deal comprehensively with developing countries' debt problems through national and international measures to make debt sustainable in the long term
   - In cooperation with the developing countries, develop decent and productive work for youth
   - In cooperation with pharmaceutical companies, provide access to affordable essential drugs in developing countries
   - In cooperation with the private sector, make available the benefits of new technologies-especially information and communications technologies

Δ Promoting Access to Credit: the Role of Women's SHGs

The exploitative landlord, the unscrupulous moneylender and the poor borrower forced to sell his land, mortgage his children or commit suicide are a pervasive feature of the landscape of rural indebtedness in India. Access to credit is necessary for a variety of reasons

- To meet seasonal fluctuations in earnings and expenditure arising from the seasonality of agricultural and non-agricultural activities.

- Credit acts as an insurance against risk for more vulnerable households, particularly in the context of an emergency / contingency or disaster.

- In the absence of effective social security programmes or public distribution systems, rural households depend on credit to meet different consumption needs, social obligations and rituals.

However, not only is the supply of formal sector or institutionalized credit to rural areas inadequate, its distribution with respect to region, class, caste and gender is highly unequal. Not surprisingly then, most poor households depend on the informal sector for loans which come at very high rates of interest, often supported by an elaborate structure of coercion, economic and non-economic (e.g. violence or sexual exploitation) as the poor have few sources of collateral. The Integrated Rural Development Programme (IRDP), one of the largest poverty alleviation programmes launched in India in the mid-1970s, sought to have the poor break out of this dependence by extending part loans and part subsidies to BPL households so that they could invest in productive activities. But IRDP, despite its target orientation failed to target the really poor. It assumed that access to credit per se would leverage change when in fact the poor needed access to a package of services. And moreover, at least in its initial phases, it was largely targeted at men, who failed to repay loans.

Why micro-credit?

The term 'micro-credit' or 'micro-finance' has gained advocacy over the last two decades partly .because of the failure of formal banking institutions to meet credit needs of the rural poor in a manner which is innovative and goes beyond loan provision to what is now acknowledged as credit plus. According to the Declaration of the Micro-Credit Summit (Washington, D.C. 1997) micro-credit programmes are those which "extend small loans to poor people for self-employment projects that generate income, allowing them to care for themselves and their families. In most cases, micro-credit projects offer a combination of services and resources to their clients in addition to credit for self-employment. These often include savings facilities, training, networking and peer support". The National Bank for Agriculture and Rural Development (NABARD) and the Reserve Bank of India (RBI) define microfinance as the "provision of thrift, credit and other financial services and products of very small amounts to the poor in rural, semi-urban or urban areas, enabling them to raise their income levels and improve living standards," Although both these definitions focus almost exclusively on income-generating opportunities, it is difficult to draw a firm line between loans for consumption and loans for productive purposes, as the credit needs of the poor are determined in a complex socioeconomic milieu (Adolph 2003).

The role of SHGs

All over the world the organisation of rural poor women (and increasingly, men) into selfhelp groups (SHGs) is widely being seen as an effective anti-poverty intervention, with a positive impact on economic growth and a range of social development indicators, including women's empowerment. While SHGs are primarily formed to provide poor women with a means of savings and access to formal credit (bank loans), today many of them are engaged in activities that go beyond microfinance, for example, 'green' agriculture, collective marketing of farm products and addressing community development issues (which include access to water, violence, infrastructure development). Many SHG leaders have also gone on to hold elected positions in their gram panchqyats, including that of the council president.

Δ Gender-based factors limiting rural women's access to credit

- Legal restrictions on credit for women-for example, many banks require the husband's or father's signature on loan applications
- Lack of information about credit availability
- Lack of security for loans in the form of land or fixed assets acceptable to lenders-typically most women pawn their jewellery to get a loan from the moneylender at exorbitant rates
- Credit provided through extension services which do not necessarily target or reach women
- Women's greater transaction costs, including distance to lenders (banks), complex procedures and constraints on mobility
- Limited credit for non-agricultural rural activities
- Lack of complementary financial and business services (a 'consultancy package') which can take into account the specific needs of different rural women rather than simply come as a 'package deal'

Δ The main objectives underlying targeting of credit to women through SHGs include:

- Breaking dependence on exploitative moneylenders by providing easy and timely availability of institutional credit at, lower interest rates.
- Organisation of women into 'solidarity groups' acts as a social guarantor (joint liability) for bank loans as most poor women do not have access to the necessary collateral (e.g. land ownership or other assets in their name).

- Small amounts of credit can help women ~cquire assets or agricultural inputs or invest in employment and income generating micro-enterprises, either collectively or as individuals.
- SHGs provide women with access to information and reduce transaction costs.
- Access to credit and investments in economic activities as well as collective mobilisation can increase women's status within the household or community.

India has one of the largest micro-finance initiatives in the world: today there are more than a million SHGs in the country with more than 17 million members (Wilson 2002). Nongovernmental organisations play a critical role in facilitating SHGs and linking them to banks (supported by NABARD's re-financing programme) or non-banking financial institutions such as BASIX, (Hyderabad). They, help in the identification of members, provide capacity building ranging from training in group building and leadership skills to financial and business management, sensitise local bank officers on working with women's SHGs and support the formation of cluster-level federations of SHGs linking them to wider market services.

Although the rapid growth and spread of SHGs in the country, the size of their total savings and the scope of activities they are engaged in are considerable, a number of questions persist particularly with respect to their impact on poverty alleviation or women's empowerment.

Δ Self-Help Affinity Groups

Affinity Groups are those "whose members are linked together by a network of relationships which enables them to perform certain traditional support functions. This affinity, which (usually) exists prior to any intervention, is based on mutual trust and reciprocity or functional support, homogeneity and voluntarism, and is adequate to cope with traditional needs. It constitutes what could be called-traditional, social and institutional capital".

Do SHG's target the poor?

Many critics feel that micro-credit has become a supply-driven industry and that banks drive NGOs to achieve targets-number of SHGs formed in a year-rather than look at who (Which women?) are members, or the pattern of loan distribution within a group (Which members borrow and for what purpose?). Typically, SHGs are small groups" of between 10 and 20 women who share relations of affinity and/or homogeneity (they come from the same caste or class background) and usually reside in the same village hamlet. Each SHG has a president, a secretary and a treasurer and usually these positions are rotated, though in areas where female literacy is low, change. Of leadership is not so common, and dependence on external accountants is not uncommon. Members meet either fortnightly or monthly to save/borrow-more frequently in the beginning during the group formation process. Amounts saved vary depending on the socioeconomic context and women's earning/saving potential; for most groups savings range from Rs 10 to Rs SO/month. These amounts are decided by the group and can increase over time if the group is functioning well. Interest rates for loans are also decided by the group and on average are about 2 per cent/month-the balance of the interest which does not go to the bank, goes towards the group account to cover internal transaction costs and to strengthen the group fund. Most SHGs will go through one to three cycles of internal loaning before they establish bank linkages.

However, in many cases SHGs tend to exclude the poorest for a number of reasons:

Social factors: The poorest are often those who are socially marginalized or excluded because of caste or community (dalits, adivasis, minorities) or physical/ mental ability; they are usually also sceptical of the potential benefits of collective action.

Economic factors: The poorest do not always have the financial resources to contribute to savings regularly, particularly during lean times (e.g. drought or after a flood). Often they cannot attend meetings if they are compelled to migrate during the lean season or work on relief sites outside the village. In some cases the group may impose a small fine on members who do not contribute the monthly amount in time-one of the SEWA-promoted groups in Kheda district, Gujarat imposes a fine of 50 paise to Re.1 if a member fails to pay her monthly installment (Shylendra 1999: 20). Such 'rules' while they may strengthen group discipline, only serve to deter the poorest. Ironically, to a question on why the poorest women in the village were not members of their SHG, the women replied: "Oh, they do not know how to save".

Intrinsic factors: These include the inherent biases of facilitating organisations who often find it more difficult to reach and motivate the poor. Some NGOs are making efforts to overcome this bias, for example, through participatory wealth ranking at the community level. But including the poorest in SHGs is not enough Programmes need to be designed so that they provide rational and sensitive investment opportunities-interest free loans as part of .a package that addresses the different loan needs of different groups of women.

Do SHGs facilitate women's economic empowerment?

In the initial years, women take loans mostly to meet household consumption needs-healthcare, purchase of food grains during times of dire need such as after a disaster, or to meet basic educational needs of children (books, pencils, school uniforms, etc.). Gradually loans are taken for more. productive activities such as the purchase of agriculture inputs, investing in new technology or livestock or starting a nonfarm-based economic activity-petty shop, cycle repair shop, tea-shop. But information about who makes decisions to take a loan or who controls the use of the loan and is responsible for its repayment is conflicting. There are some studies that suggest that many women take loans for their husbands or sons to start small businesses and that they merely become 'post-boxes' for credit and repayment, despite the assertion of 'joint control' by the women or NGO staff .

Others argue that no matter what the gender of the loanee or who controls the loan, loan taking households generally have higher income and consumption standards than non-loanee households. In many cases, women were likely to have a greater say in household decision making as well as decisions concerning their own autonomy.

Challenging gender relations: Social empowerment of women through SHGs

Most NGO staff will be quick to point out how things have changed for women who are SHG members: their greater mobility, confidence in speaking to strangers and negotiating with government officials and collective action on village issues such as alcohol brewing or sand-mining. While these are no doubt important changes for individual women and for them as a collective, there has been no radical change in, for example, the gender division of labour (sharing of the workload), or in women's access to and ownership of productive assets or in practices that discriminate against the girl child. Few women have joint bank accounts with their daughters or have created assets for them, and the practice of dowry by and large persists as does the incidence of violence, though the relationship between SHG membership and violence is complex. There is little evidence too of improvements in school enrolment and retention rates for girls; on the contrary women's engagement in credit-facilitated microenterprises may have a negative impact. Health and hygiene, food security and better nutrition are other areas that come under the quality of life, yet we have little evidence on the relationship between these social factors and the overridingly economic focus of most SHGs.

Δ Sustainability of SHG's

SHGs need continuous support to meet and deal with new challenges. There is also a need for SHGs to network among themselves to share their experiences and to develop a common strategy to face future challenges. To ensure empowerment of women through the SHG approach, projects and programmes should be designed with adequate financial and human resources to provide SHGs the required institutional support (be it from NGOs, project management, government technical agencies, financial institutions, etc) and capacity building skills (training in technical skills, basic principles of management, legal and social issues) that is required for their sustainability.

Δ Continuing support for SHGs: three reasons

- New Opportunities-It is easy to promote various SHG activities, using micro-credit in some communities because many basic needs of the community are unattended. However, the opportunity to sell goods and services in local markets will saturate soon because of the limited number of consumers and their purchasing power. Hence, there is an immediate need to identify new opportunities to sell goods and services outside the local markets.

- Capacity Building-Members of SHGs need regular training on managerial and technical aspects of the business. Linkages with technical institutions to improve the quality of the products and banking institutions for easy flow of funds are essential for further development of micro-enterprises managed by SHGs.

- Monitoring-As markets are unstable and open to competition, regular monitoring and evaluation of micro-enterprises is essential. Entrepreneurs should adopt simple systems to control finance, inventory and costs. Since many micro-enterprises are operating within a small profit margin, inadequate cost control may upset the business development. Keeping abreast of activities of large-scale manufactures, who pose a threat to SHGs is important. SHGs need to be flexible enough to make changes to their business due to competition and unfair trade practices.

(Source : Hegde and Ghorpade, 1998)

Δ Commercial development and development of micro-enterprises

The most common income generation activities that have been taken up by women SHGs in India are listed below. It is interesting to note that approximately 50% of them are non-farm. This list can be used by the DoA as a reference list to help in discussions with women farmers and other partners in identifying income generating enterprises that could be taken by SHGs in their state.

- purchase/leasing of agricultural land
- Improved seeds, bullocks, implements and other inputs for crop production
- Purchase of dairy cows, goats and chicks
- tailoring
- Paddy processing
- Rope making
- Fish-net production and repair
- mat weaving
- Production of processed milk products
- Pottery
- vermi-composting
- Nursery raising
- Bee-keeping
- Food processing (pickle and papad production and sale, curry and spice powder units)
- Leaf cup production

- Mushroom production
- Cultivation of crops
- garment production
- pasture development
- Soap powder
- Operation of flour mill
- Production of greeting cards
- Consumer store and fair price shops
- selling of cloth, sarees and bangle
- Embroidery unit
- Stationery production and sale

Δ Women empowerment through Refinement of technology

Apart from the adoption of agricultural equipment or machinery, there also exists the opportunity to improve the productivity and incomes of women farmers through the development or refinement of technologies specifically to suit their physique. Gender sensitive participatory technology development and refinement requires research scientists to understand the constraints faced by women in operating agricultural equipment and to make refinements according to feedback from women users.

A step-wise approach to refinement of technologies for women is as follows:

- Create awareness of the technology through exposure visits, village meetings, field visits etc
- Introduce the technology through demonstrations and training (not just one off activities but supported by continuous follow-up visits, advice and support).
- Collect feedback on the suitability of the technology on demonstrations days and during follow-up visits.
- Make refinements to the technology based on feedback from farmers, scientists, local blacksmiths and artisans, local manufacturers etc.
- Re-introduce the refined technology and re-test.
- If refined technology is suitable, identify backward-forward linkages to help support dissemination of the technology.

Δ Backward-forward linkages for women empowerment

Once a technology has been refined and tested in the field by women farmers and is considered to be successful, it is necessary to identify mechanisms for supporting the dissemination of that technology. Initially this requires market research. After that, a farmer support network needs to be established.

This involves examining issues including:

- Identifying manufacturing support for the technology at local level and beyond.
- Identifying possibilities for access to the technology e.g. custom hire at local level and beyond, distribution centres, service centres.
- Identifying provisions for on-field advice and support.
- Identifying provisions for maintenance and repair of the technology at local level and beyond.
- Access to credit.
- Developing linkages with line departments, policy makers, NGOs etc.

Δ Value-addition for women empowerment

The DoA has concentrated its extension efforts on promoting crop production activities. In many of the projects reviewed, there is an indication that woman farmers require activities that are not too time consuming, not drudgerous and will allow them to generate income. With this in mind, the DoA need to concentrate future efforts into developing and promoting value-addition activities by building on efforts made so far on crop production activities. In order to do this, the DoA needs to explore a number of issues including:

- Identifying quantity and quality of local produce that could be processed.
- Identifying the demand for processed products through market survey.
- Determining the cost of production through cost-benefit analysis.
- Identifying equipment required for processing produce; through purchase or hire and identifying any maintenance and repair facilities that may be needed.
- Establishing marketing methods for the processed product e.g. stalls, SHGs, village haats, selling in cities (middle-man costs) etc.
- Identifying banking and credit facilities that may be required.
- Identifying business training needs of SHGs

Δ Gender concerns in legal and policy initiatives

The government recently passed two acts which have implications for the conservation of biodiversity and the sharing of benefits arising from the use of the genetic resources and traditional knowledge. The Protection of Plant Varieties and Farmers' Rights Act (pPVFR Act, 2001) has been enacted in compliance with the TRIPs requirements for agricultural trade and has been hailed as a pioneering legislation that provides for rights and safeguards for farmers on par with plant breeders (Krishna 2004: 45).[2] While the inclusion of farmers' rights was in response to the strong stand taken by NGOs, the Act glosses over the dominant role played by women farmers "through its implicit conceptualization of the farmer as male," (Ibid. : 46). Neither in its recognition of 'community owned' varieties does it address the complex heterogeneity of communities and the embedded power relations therein. Consequently, there is an apprehension that beneflts may accrue not only to powerful elites within the community, but also to male farmers (household heads) undermining the knowledge and skills of women farmers and other marginalized groups.

- The Biological Diversity Act (BD Act, 2002) which follows a decade after the government's ratification of the CBD (1992) raise some equity concerns in the benefit sharing arrangements proposed between communities and patent claimers, which also has gender implications. For example, there is no specific, explicit requirement for obtaining prior informed consent from the owners/conservers of biological resources, one of the core principles of the CBD. Section 21 of the Act states that the National Biodiversity Authority and the proposed State Biodiversity Boards may Simply prescribe norms for equitable sharing of benefits while panchayats level Biodiversity Management Committees are to be merely consulted (MSSRF 2003). This raises serious questions of recognition of both men and women, just entitlements and the mainstreaming of gender and equity concerns in governance structures.

Δ Gender roles in agro-biodiversity

An understanding of gender issues in agro-biodiversity requires a look at the different roles and relations of men and women as part of their overall livelihood systems that comprise farms and gardens, common property resources such as pastures and forested lands, as well as protected areas. In addition to staple food production in flelds, home/kitchen gardens, largely the domain of women's responsibility, are often experimental plots where women try and adapt diverse wild plants or grown traditional subsistence food crops. The different livelihood strategies and interests, land tenure arrangements and organizational structures of different user groups (by gender, age, class, caste) as well as uneven power relations in access to, use and control over land, animal and plant resources directly influence their capacities and incentives to conserve agro-biodiversity.

Δ Global networks for food security

The MSSRF together with the FAO has played a leading role in bringing together different stakeholders to look at gender concerns in the various international and national agreements governing biodiversity, trade and intellectual property rights. The FAO has been asked to evolve a 'gender code' in the operational framework of farmers rights while the various national networks that are slowly emerging in the Asia-Pacific context have in principle, a commitment to address gender issues at all levels. The Diverse Women for Diversity (DWD), an international network of women, emerged as a Southern response to the threats to cultural and biological diversity posed by globalization. Founded in 1997, following the Conference on Plant Genetic Resources in Leipzig (june, 1996), the movement today has a presence in all continents articulating alternatives to genetically engineered food and protesting against the dumping of unsafe/untested GE products in poor, developing countries, particularly disaster prone areas, in the name of emergency relief, food aid or school mid-day meal programmes In India, the DWD articulates its commitment to diversity and food security through the National Alliance for Women's Food Rights, an umbrella forum of farmers groups, social activists and scientists concerned about the implications of the misappropriation and degradation of biodiversity for women's role in sustaining food security. Amongst the many demands outlined in the charter of the NAWFR is the call for strengthening of the public distribution system (PDS), more effective land use policies, the implementation of land reforms and the restriction or at least control of MNC investment in agriculture by the state so that farmers are not further impoverished.

Chapter 4

Rights and Privileges of Women in India

1. Constitutional Provisions

The Constitution of India not only grants equality to women but also empowers the State to adopt positive measures in favour of women for neutralizing the cumulative socio-economic, education and political disadvantages faced by them. Fundamental Rights, among others, ensures equality before the law, equal protection of law, prohibits discrimination against any citizen on grounds of religion, caste, sex or place of birth, and guarantees equality of opportunity to all citizens in matters relating to employment. Articles 14, 15, 15(3), 16, 39(a), 39(b), 39(c) and 42 of the Constitution, are of specific importance it this regard.

Constitutional Privileges

(i)   Equality before law for women (Article 14)

(ii) The Stats not to discriminate against any citizen on grounds only of religion race caste sex, place of birth or any of them {Article 15 (i)}

(iii) The State to make any special provision in favour of women and children {Article 15 (3)}

(iv) The State to direct its policy towards securing for men and women equally the right to an adequate means of livelihood [Article 39 (a)}; and equal pay for equal work for both men and women [Article 39 (d)}

(v)   The State to make provision for securing just and humane conditions of work and or maternity relief (Article 42)

(vi) The State to raise the level of nutrition and the standard of living of its people and the improvement of public health (Article 47)

(vii) To promote harmony and the spirit of common brotherhood amongst all the people of India and to renounce practices derogatory to the dignity of women {Article 51 (A)(e)}

(viii) Not less than one-third (including the number of seats reserved for women belonging to the Scheduled Castes and the Scheduled Tribes) of the number of seats to be filled by direct election in every Panchayat to be reserved for women and such seats to be allotted by rotation to different constituencies in a Panchayat {Article 243D (3)}

(ix) Not less than one-third of the total number of offices of Chairpersons in the Panchayat at each to be reserved for women (Article 243D (4))

(x) Not less than one-third (including the number of seats reserved for women belonging to the scheduled castes and the Scheduled Tribes) of the total number of seats to be filled by direct election in every Municipality to be reserved for women and such seats to be allotted by rotation to different constituencies in a Municipality {Article 243T (3)}

(xi) Reservation of offices of Chairpersons in Municipality for the Scheduled Castes, the Scheduled Tribes and women in such manner as the legislature of a State may by law provide {Article 243T (4)}

2. Legislative Provisions

Legal Rights

To uphold the Constitutional mandate, the State has enacted various legislative measures intended to ensure equal rights, to counter social discrimination and various forms of violence and atrocities and to provide support services specially to working women.

Although women may be victims of any of the crimes such as.Murder., Robbery, Cheating etc. the crimes, which are directed specifically against women, are characterized as.Crime Against Women.. These are broadly classified under two categories

(1) The Crimes Identified Under the Indian Penal Code (IPC)

(i)   Rape (See. 376 IPC)

(ii) Kidnapping & Abduction for different purposes (See. 363-373)

(iii) Homicide for Dowry. Dowry Deaths or their attempts (See. 302l304-B IPC)

(iv) Torture, both mental and physical (Sec. 498-A IPC)

(v)   Molestation (See. 3541 PC)

(vi) Sexual Harassment (See. 509 IPC)

(2) The Crimes identified under the Special Laws (SLL)

Although all laws are not gender specific, the provision of law affecting women significantly have been reviewed periodically and amendments carried out to keep pace with the emerging requirements.

Some acts which have special provisions to safeguard women and their interests are

(i)   The Employees state Insurance Act, 1948

(ii) The Plantation Labour Act, 1951

(iii) The Family Courts Act. 1954

(iv) The Special Marriage Act, 1954

(v) The Hindu Marriage Act, 1955

(vi) The Hindu Succession Act, 1956

(vii) Immoral Traffic (Prevention) Act, 1956

(viii) The Maternity Benefit Act, 1961 (Amended Act, 1995) (ix) Dowry Prohibition Act, 1961

(x) The Medical Termination of Pregnancy Act. 1971

(xi) The Contract Labour (Regulation and Abolition) Act, 1976 (xii) The Equal Remuneration Act, 1976

(xiii) The Child Marriage Restraint (Amendment) Act, 1979

(xiv) The Criminal Law (Amendment) Act, 1983

(xv) The Factories (Amendment) Act, 1986

(xvi) Indecent Representation of Women (Prohibition) Act, 1986 (xvii) Commission of Sati (Prevention) Act, 1987

3. Special Initiatives for Women

(i) National Commission for Women

In January 1992, the Government set-up this statutory body with a specific mandate to study and monitor all matters relating to the constitutional and legal safeguards provided for women, review the existing legislation to suggest amendments wherever necessary etc.

(ii) Reservation for Women in Local Self-Government

The 73rd Constitutional Amendment Acts passed in 1992 by Parliament ensure one third of the total seats for women in all elected offices in local bodies whether in rural areas or urban areas.

(iii) The National Plan of Action for the Girl Child (1991-2000 AD)

The plan of action is to ensure survival, protection and development of the girl child with the ultimate objective of building up a better future for the girl child.

(iv) National Policy for the Empowerment of Women, 2001

The Department of Women & Child Development in the Ministry of Human Resource Development has prepared a National Policy for the Empowerment of Women. in the year 2001. The goal of this policy it to bring about the advancement, development and empowerment of women.

Chapter 5

Impact of Liberalization
on Agricultural Workers and Women

There is apprehension that economic liberalization, which tends to induce privatization and market-led technological change, may affect employment and income prospects of rural women adversely. For instance the impact of opening of fisheries and agro processing to multinational and corporate reportedly displaced million of workers.

The impact of economic liberalization on agricultural workers and women may be summarized as follows:

- The volatility of international prices of agricultural commodities has affected agricultural workers in India adversely. Particularly during the last few years or so, most of India.s agricultural commodities have lost international competitiveness due to low international prices. The exports of agricultural commodities have fallen. Simultaneaously, there have been increased imports of edible oils and raw cotton which have depressed the domestic prices of these commodities.

- It is apprehended that the process of economic liberalization will ultimately result in a situation in which multinational companies will dominate the agricultural scene in India and small farmers may lose their grips over agriculture, posing a risk of livelihood insecurity for millions of agricultural workers and rural women.

- In the wake of WTO, not only agricultural exports have declined, but also rate of growth of agricultural output and employment. Consequently, both cultivating households as well as landless particularly women labourers have suffered a loss of income.

- In the wake of economic liberalization, cultivators as well as agricultural labourers look depressesed because of deterioration in their income levels as well as uncertain future. Also the gap between agricultural and nonagricultural workers is widening.

- There is a secular worsening of the condition of agricultural labourers due to increase actualization of labour force and inability of the organizes farm and non-farm sectors to absorb the growing labour force.
- In the plantation sector too, the decline in the export earnings from tea, coffee and rubber and low domestic prices have affected both income and employment of women. Particularly, smaller plantations find it difficult to maintain the levels of employment and wages of the labourers.

Δ Women as Knowledge Keepers and Custodians of Cultural Diversity

Nature has given us abundance; women's indigenous knowledge of biodiversity, agriculture and nutrition has built on that abundance to create more from less, to create growth through sharing.

- Women farmers share seeds freely and with sharing as a base, there can never be scarcity.
- The giving and sharing of food in abundance has been the basis of ensuring food security.
- The giving and sharing of knowledge has been the basis of the growth and evolution of knowledge. An economy of sharing is an economy of abundance.
- This worldview of abundance is based on sharing and on a deep awareness of human as members of the earth family. This awareness that in impoverishing other beings, we nourish ourselves is part of our present ecological knowledge and ancient wisdom.
- Without giving and sharing there can be no sustainability; with our sustainability, there can be no space.
- Diverse Women for Diversity (DWD) movement celebrates sharing for sustainability and peace through food festivals, exposure tours, training programmes in biodiversity conservation, sustainable agriculture, indigenous healing systems, water conservation, capacity building for leadership and good governance.

Δ Vision for the XII Five Year Plan

The vision for the XII Five Year Plan is to ensure improving the position and condition of women by addressing structural and institutional barriers as well as strengthening gender mainstreaming.

Δ Goals for the XII Five Year Plan

- Creating greater 'freedom' and 'choice' for women by generating awareness and creating institutional mechanisms to help women question prevalent "patriarchal" beliefs that are detrimental to their empowerment.

- Improving health and education indicators for women like maternal mortality, infant mortality, nutrition levels, enrolment and retention in primary, secondary and higher education.

- Reducing the incidence of violence against women and providing quality care services to the victims.

- Improving employability of women, work participation rates especially in the organised sector and increased ownership of assets and control over resources.

- Increasing women's access to public services and programmes through establishing and strengthening convergence mechanisms at multiple levels, creation of physical infrastructure for women and improving the capacity of women's organizations and collectives.

- Ensuring that the specific concerns of single and disadvantaged women are addressed.

Δ Institutional Arrangements for women empowerment

- Strengthening of National Women's Machineries is vital to achieving women's empowerment. This includes the MWCD as the nodal Ministry and its attached and autonomous organisations, National Commission for Women, Rashtriya Mahila Kosh and the National Mission for Empowerment of Women (NMEW). The important role of MWCD is to facilitate mainstreaming of gender concerns in policies, programmes and schemes of all Ministries and Departments, to implement special legislations and welfare programmes and Schemes for women as well as to undertake advocacy. To enable the Ministry to play this role effectively, its women's wing will sought to be strengthened in the XII Plan.

- The mandate of the National Commission for Women is to protect and safeguard the rights of women. The activities of the Commission include receiving complaints, undertaking suo moto enquires in cases of deprivation of rights of women, conducting Parivarik Lok Adalats and legal awareness programmes and organising public hearings. The strengthening of the Commission, in terms of staff requirement, is underway and will be completed.

- Interaction between the NCW and SCWs needs to be enhanced so that the monitoring of legal safeguards to women is harmonised across the country. NCWs linkage with the State Commission would be looked into in the XII Plan. For this the possibility of teleconference linkage and assistance for awareness generation by the National Commission to the State Commissions will be explored.

- Gender being a cross cutting issue, various Ministries/Departments have been undertaking measures for the empowerment of women. Convergence of these programmes and schemes is essential to ensure that their benefits are effectively accessed by women. With the specific objective of ensuring convergence and better coordination among the schemes/programmes of various Ministries/Departments, the Ministry launched the National Mission for Empowerment of Women. The Mission will be fully operationalised. Its role would be to provide a strong impetus for reform by catalyzing the existing system, ensuring better coordination and convergence of all development programmes impacting women in close collaboration with grassroots structures and enabling participatory approaches and processes. The Mission would work to achieve convergence at all levels of governance. It would have an overarching role in promotion of women's issues across economic, social, legal and political arena. This would, inter alia, include generating awareness; building strategies to question prevalent "patriarchal" beliefs; establishing a convergence mechanism at multiple levels; formation women's collectives and improving their capacity to access the benefits of government schemes, programmes, laws and policies; and developing empowerment indicators relating to the survival, visibility, freedom and equality of women.

- The Rashtriya Mahila Kosh (RMK) as the credit extending arm of the MWCD will be strengthened. It will be restructured as a non-banking finance company of systemic importance with an enlarged corpus of Rs. 500 crores. This will enable it to reach out to a larger number of poor, assetless and marginalised women for income generating, production, skill development and housing activities.

- To enable all institutions to identify and respond to gender issues, the XII Plan will push for establishing Gender Focal Points within various organisations like the Ministries/Departments of the Central Government and Urban and Rural Local Bodies.

- The process of engendering institutions would require that National Women's Machineries are engaged in a gender analysis of not only programmes and projects but also of institutions like the Panchayati Raj institutions (PRIs,) judiciary, the enforcement machineries etc.

Chapter 6

Women Entrepreneurship and Its Role in Economic Development

Introduction

An entrepreneur is someone who is a risk taker and is ready to face challenges. Women entrepreneur is a woman or group of women who initiate, organize and run business enterprise."

Entrepreneurship refers to the act of setting up a new business or reviving an existing business so as to take advantages from new opportunities.

Thus, entrepreneurs shape the economy by creating new wealth and new jobs and by inventing new products and services.

However, an insight study reveals that it is not about making money, having the greatest ideas, knowing the best sales pitch, applying the best marketing strategy

It is in reality an attitude to create something new and an activity which creates value in the entire social eco-system. It is the psyche makeup of a person

It is a state of mind, which develops naturally, based on his/ her surrounding and experiences, which makes him/ her think about life and career in a given way.

It is a state of mind, which develops naturally, based on her surrounding and experiences, which makes her think about life and career in a given way.

The women have achieved immense development in their state of mind.

With increase in dependency on service sector, many entrepreneurial opportunities especially for women have been created where they can excel their skills with maintaining balance in their life. Accordingly, during the last two decades, increasing numbers of Indian women have entered the field of entrepreneurship and also they are gradually changing the face of business of today, both literally and figuratively. But still they have not capitalized their potential in India the way it should be.

Importance

Women entrepreneurship has been recognised as an important source of economic growth. Women entrepreneurs create new jobs for themselves and others and also provide society with different solutions to management, organisation and business problems. However, they still represent a minority of all entrepreneurs. Women entrepreneurs often face gender-based barriers to starting and growing their businesses, like discriminatory property, matrimonial and inheritance laws and/or cultural practices; lack of access to formal finance mechanisms; limited mobility and access to information and networks, etc.

Woman constitutes the family, which leads to society and Nation. Social and economic development of women is necessary for overall economic development of any society or a country. Entrepreneurship is the state of mind which every woman has in her but has not been capitalized in India in way in which it should be. Due to change in environment, now people are more comfortable to accept leading role of women in our society, though there are some exceptions.

Our increasing dependency on service sector has created many entrepreneurial opportunities especially for women where they can excel their skills with maintaining balance in their life.

Reasons for Women Opting For Entrepreneurship

Self determination, expectation for recognition, self esteem and career goal are the key drivers for taking up entrepreneurship by women (Moore & Buttner, 1997).

Sometimes, women chose such career path for discovering their inner potential, caliber in order to achieve self satisfaction.

It can also provide a mean to make best use of their leisure hours. However, dismal economic conditions of the women arising out of unemployment in the family and divorce can compel women into entrepreneurial activities.

Obstacles for Women Entrepreneurship

Financing

Balancing of life

Lack of family support for the women

The other hindering External Factors include

Gender discrimination,

Inaccessibility to information

Training opportunities

Infrastructure etc.

Some Internal Factors like risk

Aversion by women

Lack of confidence

Lack of vision of strategic leader etc.

Tips for Women Entrepreneurs
- Start a business that works for you and your personal life
- Research the product/ service
- Assess the market
- Start business with adequate funds
- Do networking
- Consult with professionals

What women entrepreneurs need most is
Access to information

Women are often not connected to the business networks through which information flows and contacts and connections are shared.

When women find a way to share their know-how and their skills, their businesses benefit

Δ Access to markets

Finding markets for the goods women produce is a challenge,

Connecting women entrepreneurs with interested customers provides a long-term boost to the women's businesses, with the possibility of creating long-term sustainability.

Δ Access to capital

Most of women have little in terms of collateral or a prior track record of business achievement, both of which are critical parts of securing a bank loan.

Tools such as loan guarantees and women-focused financial products can help women get around the barriers presented by a lack of available investment capital

Microfinance is another excellent first step for many, with microloans providing the funds to get a small venture started.

Policies and Schemes for Women entrepreneurs in India

In India, the Micro, Small & Medium Enterprises development organisations, various State Small Industries Development Corporations, the Nationalised banks and even NGOs are conducting various programmes including Entrepreneurship Development Programmes (EDPs) to cater to the needs of potential women entrepreneurs, who may not have adequate educational background and skills. The Office of DC (MSME) has also opened a Women Cell to provide coordination and assistance to women entrepreneurs.There are also several other schemes of the government at central and state level, which provide assistance for setting up training-cum-income generating activities for needy women to make them economically independent. Small Industries Development Bank of India (SIDBI) has also been implementing special schemes for women entrepreneurs.

In addition to the special schemes for women entrepreneurs, various government schemes for MSMEs also provide certain special incentives and concessions for women entrepreneurs. For instance, under Prime Minister's Rozgar Yojana (PMRY), preference is given to women beneficiaries. The government has also made several relaxations for women to facilitate the participation of women beneficiaries in this scheme. MSE Cluster Development Programme by Ministry of MSME, the contribution from the Ministry of MSME varies between 30-80% of the total project in case of hard intervention, but in the case of clusters owned and managed by women entrepreneurs, contribution of the M/o MSME could be up to 90% of the project cost. Similarly, under the Credit Guarantee Fund Scheme for Micro and Small Enterprises, the guarantee cover is generally available up to 75% of the loans extended; however the extent of guarantee

Some of the special schemes for women entrepreneurs implemented by the government bodies and allied institutions are provided below.

Schemes of Ministry of MSME

- Trade related entrepreneurship assistance and development (TREAD) scheme for women
- Mahila Coir Yojana

Schemes of Ministry of Women and Child

Women's Role in Economic Contribution

In line with the improvement of women's education, women are no longer the minority in fields that were dominated by men in the past. The field of information technology creates many opportunities for the development of women's talents in this specific field.

The increase in the number of women who lead their own business, especially the ones in small and medium scale enterprises.

Women's leadership is able to gain high loyalty due to the fact that they are the ones that are able to conduct clean, ethical, transparent and honest management.

The increase in the number of women who lead their own business, especially the ones in small and medium scale enterprises.

Economic development is a process through which overtime sustained increase occur in nation's per capita real income accompanied by significant structural changes that allow for elevated income distribution and large increase in individual economic well-being.

Rise in income must be evident through such changes in basic living conditions as improved nutrition and high nutritional and clothing standards, improved housing, improved health and health care, low infant mortality rate and higher literacy rate

Role of women entrepreneurship in economic development

- Economic development refers to Qualitative measure of progress in an economy, refers to development and adoption of new technologies, transition from agriculture based to industry based economy, and general improvement in living standards

Overholt (1986) pointed out that the role of women entrepreneurship in development

Women's contribution brings economy growth and development.

When women are fully involved, the benefits can be seen immediately

Families are healthier and better fed, their income, savings and reinvestment go up.

What is true of families is also true of communities and, in the long run, of whole countries

Women are the Third World's powerhouse

They produce a staggering 60 percent of all food

Run 70 percent of small-scale businesses an

Make up a third of the official labour force

By any measure - income, education, health, land ownership, legal rights or political power

Women get a raw deal in education, health services, equal pay employment, access to land and finance. It is becoming increasingly clear

However, that there will only be sustainable development in the Third World when women play an equal part in decision making

Conclusion

Enterprise births in local areas and facilitation of the development of new women-owned firms can have positive impacts on

Job creations,

Productivity growth,

Tax revenues

The availability of goods and services and

The provision of positive role models.

Entrepreneurship among women is critically important for local development and for economic development.

Women are the economy driving force

Combination influence of more education, technology and fast economic growth make women more assertive concerning their right, more aggressive in reaching their ambition while we already acknowledged that the number of women in the work force from country to country are almost as high as those of men.

Their contribution in providing job openings in business sectors continues to rise. They are involved in enterprises significantly in the levels as managers, entrepreneurs, owners and investor.

Chapter 7

Women Empowerment in Dairy Sector

- The overall farming performance is the farming productivity behaviour of farm women comprising six components viz. cropping intensity, cropping yield index, milk yield index, level of adoption, commercialization index and expansion/diversification index in a rice based farming system. The result indicate that there is significant difference between small, medium and large farms women with respect to overall farming performance. It is observed that small farm women had less cropping intensity, crop yield index, level of adoption, commercialization and expansion/diversification index when compared to medium and large women. However, in case of milk yield index, small farmwomen are on par with medium but higher than the large farmwomen. (Reddy 2003).

- According to Verma (1992), Animal Husbandry is predominantly a male affair in case of high, economic status as majority of them employ, permanent male labour to look after the animals, whereas it is predominantly a female affair in case of farmers of medium and low socio-economic status. On an average, a woman devotes 3.5 hours per day for animal husbandry activities against only 1.6 hours per day devoted by men in this category.

- Women accounted for 93% of total employment in dairy production. Depending upon the economic status, women perform the tasks of collecting fodder, collecting and processing dung. Women undertake dung composting and carrying to the fields. Women also prepare cooking fuel by mixing dung with twigs and crop residues. Though women playa significant role in livestock management and production, womens control over livestock and its products is negligible. Men, leaving only 14% to women, assume the vast majority of the dairy cooperative membership.

- In tribal communities in low rainfall, semi-arid and arid areas much of the work with regard to animal management has to be looked after by women due to migration of males for work. However, in many cases the income from dairy animals does not remain in the hands of women and neither does the decision regarding sale and purchase. However, due to the move to develop women's dairy cooperatives in many states in India women have better control over sale of milk and use of income from it. Another positive development is recognition of women as members of dairy co-operative societies, so that the price of milk supplied to the society can be paid to the women directly. Till a few years ago women were not made registered members of the dairy co-operative society (the registration was in the name of the husband and thus he collected the money for milk produced and supplied by the women).

- Women are well aware of each animals behaviour and production characteristics. Women are knowledgeable about local feed resources and are able to identify beneficial grasses, weeds and fodder tress for feeding of dairy animals. While the contribution of women to the animal's management is recognized, the experiences of women regarding animal diseases and their perceptions are ignored. Now there is some realization about the knowledge possessed by women and the need to improve their knowledge, skills and awareness.

Feeding: Participation of women in indoor feeding activities such as providing water to animals, mixing ration and preparing feed is very high. As far as the involvement in outdoors activities is concerned, such as cutting and transportation of fodder, the deployment of women is low.

Health care: The share of rural women is found more in all animal health care related activities performed at home. Activities like care of sick animals, care of animals during pregnancy, care of newly born calf and applying and giving medicine at home are performed jointly by women members of the family.

Processing and marketing related activities: The participation of rural women in processing activities is found higher than in marketing related activities such as purchase of animals, selling of milk, disposal of animals etc.

Management: Women perform all the activities related to management of animals like cleaning of animals and their shed, milking and making cow dung cakes.

- In India, Animal Husbandry is a vocation for millions of small farmers and land less people, a large percentage of them raise animals on crop residues and common property resources. When in some developed countries, less than 3% of the adult population is engaged in agriculture and animal husbandry, 70 percent of Indian population is dependent on agriculture and animal husbandry, of which 30 percent are landless. Women provide 60 percent of the livestock farming labour and more than 90 percent of work related to the care of animals is rendered by women folk of the family.

- During 2001-02, the contribution of livestock sector to the country's GDP was 5.59 percent. The value of output of livestock and fisheries sector was estimated to be around 179544 crores, which is about 27.7 percent of the total value of output of Rs.648122 crores from agriculture and allied sectors. With an estimated output of Rs. 103804 crores, the contribution of milk was higher than paddy (Rs. 73965 crores), wheat (Rs. 43816 crores) and sugarcane (Rs. 28592 crores). Milk and livestock sector provides regular employment to about 11 million in principal status and 8 million in subsidiary status. Women in animal husbandry and livestock sector constitute 69 percent of the labour force as against 35 percent in crop farming (Economic Survey 2002-03).

- Dairy farming has grown from largely unorganized, complex activity into a vast organized industry that encompasses not only increased production of milk and milk products, but also the breeding of higher yielding cattle, and the scientific rearing of cattle and production of cattle feeds. This has proportionately increased the demand of trained manpower in this sector. However, the implementations of WTO policies will totally ruin the dairy sector and thus affecting the available employment opportunities.

- A common feature in region where dairying is an important commercial activity of the rural population as in Gujarat and Punjab, is that it is the women folk who essentially maintain and manage the dairy cows and buffaloes. In areas, where the milk produced is channelled through dairy plants, bringing daily or weekly income to the household, the dairy activity raises the status of women.

- Winner of the prestigious Magsaysay Award, Mrs. Ela Bhatt, highlights the need for giving women their due place in dairy development. For their empowerment and economic well being. Women's access to training in modern dairying and cooperative management is essential.

- The membership in most of India's 70,000 village level dairy cooperative societies (DCS) is heavily dominated by men. The picture is now gradually changing in the favour of women. Efforts are on to give them their due place in dairy development.

- Presently, some 2,476 all-women DCS are functioning in the country in selected States. Out of 9.2 million total memberships in DCS 1.63 million are women (18 per cent). However, women constitute less than three per cent of total board member.

Δ Women Second in the Land Agenda

- The land reform measures adopted by the Indian Union or respective states are not in accordance with any personal law of any community. In matter of women.s right to inheritance of agriculture land, states either have their own policy or refer it to the principle of personal laws. (Jyoti Gupta 2002)

- In a discussion conducted by in West Bengal the women asserted the importance of ownership of productive resources in their name. Their first demand was,.We want land in our own name. Various reasons were forwarded by the women for such a demand. (Jyoti Gupta 2002).

1. Old age security if sons do not look after them.

2. The predominant notion that their daughters can inherit land if the mother has it in her own name.

3. Women can use the property to pay for their daughters. marriage in the absence of the husband; as sons usually inherit the property and may be unwilling to pay for their sister's marriage.

4. Lack of security in the matrimonial home, as women do not have any legal right to property, be it land, cattle or house.

5. Given the rise in the cases of divorce, desertion and physical violence against Women by husbands, the women expressed the urgent need for secure shelter, be it legal right to be homestead of their parents or land to cultivation.

6. The younger women especially those who have already divorced, deserted or remained Unmarried, preferred to be given a share in he parents properly, be it land, house or both.

7. Women felt that they should have independent access to ownership of productive resources to be able to counter the system of dowry, as well as to free themselves of a dependent status all their lives.

8. The women also pointed out that apart from ploughing, they participate in all agricultural operations. Women who belonged-to peasant house hold, but did not have to work as agricultural labourers also pointed out that they bear all responsibility for the agriculture produce once it is brought home from the field, yet they are not considered agriculturist or cultivation only their husbands are recorded as cultivators. Women felt that ownership in their name would make a difference to their access and control over the women also told that though are capable of ploughing but they are not allowed.

9. While women do not have ownership rights recorded in their name, they however bear the responsibility of returning debts, incurred by their husbands in lieu of land mortgage.

10. Women are often not informed about such debts or transactions the husbands enters into with the assets of the family. The women felt that such a solution could be checked if they as owners were to be signatories to the transaction.

Chapter 8

Women Empowerment
Programmes and Schemes

Δ Programmes for Rural Women-A Brief Overview
Programmes with Specific Focus on Farm Women

The first major project to address the training and extension needs of woman farmers in India was The Women and Youth Training and Extension Project (WYTEP) implemented by the Department of Agriculture, Government of Karnataka, with funding from the Danish International Development Assistance (DANIDA) in 1982. Since then, DANIDA has funded three more projects in another three states, Tamil Nadu, (Tamil Nadu Women in Agriculture Project, TANWA), Orissa (Training and Extension for Women in Agriculture Project, TEWA) and Madhya Pradesh (Madhya Pradesh Women in Agriculture Project, MAPWA):

Δ Women Youth Training and Extension Project (WYTEP), Karnataka

WYTEP was implemented by the Department of Agriculture, Government of Karnataka, with funding from the Danish International Development Assistance (Danida) in 1983. The project is now in its third phase of implementation with a development objective of securing the utilization of women's potential in agricultural production on small and marginal holdings for the betterment of quality of life for all the members of the family.

Phase I and II

The project focused on activities specific to woman farmers including:
* Selection of crop and variety
* Seed selection and treatment
* sowing and transplanting

- Application of fertilizer both during sowing and as top dressing
- weed control, integrated pest management and disease control
- Harvesting and post-harvest activities including processing, storage, etc,

The main components in the first two phases were:

- Training women farmers in technologies relevant to their role in agriculture-for this, training centres were established and equipped to provide residential training. There are now 16 training centres in the state. The project has also collaborated with other allied government line departments in providing training in sericulture, animal husbandry and horticulture activities, to widen the scope of livelihood options and opportunities available to women farmers.
- Providing extension services to women farmers through activities such as pre-seasonal camps, demonstrations done at the village level and providing regular support to them.
- Organizing women farmers into groups to enable them to learn from each other, take up collective action in procuring inputs for agriculture and allied activities. Groups were strengthened through group leadership which was developed through link worker training. Eligible groups were also provided seed money to promote savings and credit activities.

The main components of Phase III are:

- Mainstreaming Gender in Agriculture Extension. The general extension system (GES) and its field extension team (FET) take on the responsibility for woman farmers extension services. Prior to this, the extension services for woman farmers was provided largely by women officers of the DoA. It is hoped that, women farmers are able to reach and use the services that the extension service provides. Further more, the Agriculture Department is taking up extensive staff development activities to make the system gender responsive and to keep the needs of women farmers at the forefront of planning and implementing agricultural programmes.
- Developing and disseminating relevant technologies specifically for women. In order to pay attention to the technological problems faced by women in small and marginal holdings, funds will be provided at both the District and State levels (to be managed by the District level Technical Advisory Committees and the State level Technical Advisory body) to develop and disseminate technologies specifically for women.
- Providing gram panchayat funding for women farmers. The main objective here is to provide opportunities and funds to gram panchayats so that they develop an interest in the extension activities of women farmers. This is envisaged as an effort in 'engendering the GP budget' in a critical sector like agriculture at the most basic unit of decentralized democracy. This will be piloted in a few GPs in the first year before expansion to more GPs.

Δ Tamil Nadu Women in Agriculture (TANWA), Tamil Nadu

The Project has been operational since October 1986. The project was initially implemented in six districts of Tamil Nadu and then expanded into ten districts. Phase II of this Project is was initiated in 1993 covering all the districts (except Chennai) and would end in March 2003.

The main objectives of the project were:

- To increase agricultural productivity and to improve the economic and food security of small and marginal women farmers.
- enable women farmers to choose and adopt relevant agricultural technologies and practices.
- To disseminate agricultural knowledge and skills from women farmers to fellow women farmers.
- To improve women's access to and improve their ability to use existing agricultural services.

Δ Village based training programmes were also planned and conducted on the following aspects:

- Increasing fertiliser use efficiency
- Organic manure conservation (FYM, compost enriched FYM)
- Use of bio-fertiliser
- Tree cropping
- Field identification of pest/diseases
- Seed treatment
- Rat control
- Grain storage
- Crop wise fertiliser application and plant protection measures
- Communication skills

After the training, follow-up visits were made (ten visits spread over two years) by women agricultural officers to provide guidance and support to the trained woman farmers on how to adopt the skills learnt during the village level training.

Specialised training programmes were also organised for 1-2 days in other topics related to agriculture, for example, animal husbandry, agro-forestry, cash crops, sericulture, and pisciculture.

Trained women farmers were also encouraged to establish women's groups in their villages to share their knowledge and experiences with others.

Δ Training and Extension for Women in Agriculture (TEWA), Orissa

The TEWA project (funded by DANIDA) was implemented in Orissa in 1987 by the Department of Agriculture (Government of Orissa). The first phase of the project covered four districts and was completed in 1995. The second phase covering an additional four districts commenced in 1995 and is expected to continue till March 2003.

The target group under the TEWA project were the farm families with small and marginal land holdings who were actively involved in agriculture activities. Lady village agricultural workers (LVAWs) after having worked with women farmers for two years, enter into the general extension system to provide extension and capacity building support to the target group.

Through training, extension and field visits to women farmers, the TEWA project has concentrated on:

- Creating awareness of increasing agricultural production
- Improving accessibility of local agricultural/allied extension staff to woman farmers and vice versa.
- Improving the adaptability of messages on seed testing, seed treatment, planting of paddy seedlings, maintenance of plant population measures, raising backyard kitchen gardens.
- Enhancing the adaptability of new messages on the use of bacteria culture in pulses and oilseeds, rodent control for storage of grains, soil health management practices, rainfed farming technologies, water management, integrated nutrient management, integrated pest management and post harvest technology.
- Preparation and use of bio-fertilisers
- Crop diversification experiments such as sunflower, off season vegetables, flower and fruit cultivation, mushroom cultivation, dairying, poultry, fishery and sericulture.

Δ Madhya Pradesh Women in Agriculture (MAPWA), Madhya Pradesh

MAPWA is a skill oriented agriculture training and extension project for woman farmers which was launched in 9 districts of Madhya Pradesh in 1993. The second phase of the project was initiated in 2002 covering all the remaining districts. The main foci of MAPWA were:

- Village based training for women farmers by a team of 2 agricultural development officers (for farm women training) during the lean season followed by 3 follow-up visits per year over a duration of 2 years. These visits were organised after 20-25 day intervals subsequent to the completion of village based training.
- Specialised training on income generation activities in agriculture and allied agriculture, for example, poultry rearing, mushroom production

and horticulture for the women farmers who successfully completed village based training.

- In Gujarat and Andhra Pradesh similar programmes have been under implementation with funding assistance from the Government of Netherlands. These are: Training of Women in Agriculture Project (TWA) in Gujarat and the Training of Women in Agriculture Project (ANTWA) in Andhra Pradesh. The respective Departments of Agriculture are responsible for their implementation.

Δ Training of Women in Agriculture (TWA), Gujarat

The TWA project was a bilateral funded project by the Dutch Government and implemented by the Department of Agriculture through farmer training centres (FTCs) in selected districts. Phase I of the project was launched in 1989 in six districts of Gujarat. This was followed by phase II in 1997 which covered an additional six districts and is expected to end in 2003.

Five day training programmes were organised by the FTCs each year for about 25 women farmers selected from a cluster of 2-3 villages. Village based training was also organised at the village level for those women farmers who live far away from the FTCs. Specialised training of 7-10 days duration was organised each year for those who were interested in agro-based enterprises to generate additional income. The specialised training was given in dairying, vegetable cultivation, animal husbandry and horticulture.

The trained women farmers along with untrained women farmers from the villages were formed into charcha mandals. The FTC organised three days training courses each year for about 25 leaders of charcha mandals. Pre-seasonal follow-up training camps were also organised by the FTCs at the village level during both kharif and rabi season.

Δ Training of Women in Agriculture Project (ANTWA) in Andhra Pradesh

The ANTWA project was initiated in 1994 to acquaint women farmers who had small and marginal land holdings with improved technological skills in agriculture through a training cum extension program. The project covered six districts in Phase I and was completed in 2001. Phase II was initiated in 2001 and covered an additional six districts in partnership with Andhra Pradesh Women's' Cooperative Finance Corporation (APWCFC).

The main objectives of the project were:

- To train women farmers with small and marginal landholdings with the latest agricultural technologies and practices which were relevant to their farming systems.
- To improve the ability of the trained women farmers to utilise the existing agriculture extension services

- To develop and increase knowledge of women farmers through agricultural training and providing agricultural extension service support.

These objectives were to be met through the two major components of the program; training and extension. Training was done at the Farmers Training Centres (FTCs) and extension support is provided by Agriculture Officers specially posted in the sub-division for this project.

Village based training based on locally relevant, low or no cost technologies, was the first major activity. FTC staff followed up the training by visits to the villages and organizing pre-seasonal training to discuss specific requirements of the cropping season.

For those who wanted to gain more in-depth information on a particular subject, specialised training was provided. A study tour within the state was also arranged for trained woman farmers to provide them with exposure to improved farming techniques.

Δ Central Sector Scheme of Women in Agriculture (CSSWA)

The central sector scheme on Women in Agriculture was launched on a pilot basis under the Government of India's Eighth Plan. This was done in one district each in 7 states (Punjab, Haryana, Uttar Pradesh, Himachal Pradesh, Maharasthra, Kerala and Rajasthan). The project was extended during the Ninth Plan to cover one district in each of 8 North-Eastern Hill States (Manipur, Meghalaya, Mizoram, Arunachal Pradesh, Nagaland, Tripura, Assam and Sikkim).

• UNDP-Government of India sub programme on food security

Under the United Nations Development Programme-Government of India sub programme on food security, four programmes specifically focussing on women in agriculture, have been implemented since 1999. Table A2 provides a summary of the major features of these programmes.

Other programmes for woman farmers include the "Women's Training Programme" under World Bank assisted Agricultural Development Project (ADP). The programme was launched in 1992 to cover 14 districts of the state of Rajasthan. More recently, the state of Uttaranchal has initiated a project on Women in Agriculture in 68 blocks of the state covering 340 women groups and the activities include training, study tours, exhibition of products, demonstrations and seminars.

Δ Programmes with Women in Agricultural Communities as an Important Focus

Programmes and projects that have a broader development agenda but contain a significant component on women in agriculture have been under implementation in India. For example, the DFID funded Eastern India Rainfed Farming Project, the IFAD funded Tamil Nadu Women Development Project, the Uttar Pradesh Sodic Land Reclamation Project, the "Kudumbashree" programme supported by the State Poverty Eradication Mission, Kerala) and the NDDB supported Women Co-operative Dairy Programme.

Δ Other Rural Women's Development and Empowerment Programmes

A number of programmes have been implemented by other government departments, such as, the Department of Women and Child Development, which complement and supplement the programmes offered by the state departments of agriculture. The major programmes are:

Δ Indira Mahila Yojana (IMY)

The IMY project was launched in 1995 and has been replaced by the Swayamsidha scheme in 2001. The scheme was implemented in 238 blocks in the country with the aim of creating self-help groups (SHGs) for the empowerment of women. The main strategy of the scheme was to create an organisational base for women so that they could come together and discuss their needs and requirements from the existing departmental programmes of the State and Central Governments. By 1999, 40,000 SHGs had been formed under the scheme.

The programme also envisaged the development of Indira Mahila Kendras (IMKs) at the Anganwadi level and integrating or associating with other groups formed under other programmes, such as, health and adult literacy to provide the grassroots level organisation for women. It has been proved by several experiments in different parts of the country that women SHGs become a very strong medium for enabling access to information, knowledge and resources.

Δ Swayamsidha (IWEP)

The Swayamsidha, is an integrated scheme for women's empowerment (IWEP) which was launched in 2001 and replaced the IMY scheme. The IWEP programme was based on the SHG philosophy but was also aimed at empowerment of women through generating awareness, economic empowerment and convergence of various schemes.

The vision of the IWEP was to empower women so that they would

- demand their rights from their family, community and government
- have increased access to and control over, material, social and political resources
- have enhanced awareness and improved skills
- be able to raise issues of common concern through mobilization and networking

The IWEP created approximately 53000 SHGs, approximately 26500 Village Societies and 650 Block Societies which has benefited about 930 000 women so far.

Δ Rural Women's Development and Empowerment (Swa-Shakti) Project

The Rural Women's Development and Empowerment Project, also known as the Swa-Shakti Project, was sanctioned in 1998 as a centrally sponsored project for a period of 5 years. The overall objective of the project was to strengthen the processes involved in the creation of an environment to empower women. The specific objectives were to:

- establish 12000 self-reliant women's Self-Help-Groups (SHGs) (each having 15-20 members) to improve the quality of women lives via enabling greater access to, and control over resources
- sensitise and strengthen support agencies institutional capacities to pro-actively address women's needs
- develop linkages between SHGs and lending institutions to ensure women's continued access to credit facilities for income generation activities
- enhance women's access to resources to reduce drudgery and enable a better quality of life
- increase control over income and spending, through involvement in income generating Activities.

Δ Distance Education Programme

SHGs have emerged as one of the major strategies to empower women. However, the experience of various schemes has shown that the sustainability of the majority of the SHGs was a problem and one of the major reasons was a lack of proper training. Incomplete or ineffective training in the formation of SHGs meant that the full potential of women's groups could not be realised. The vast geographical canvas also impeded timely transmission of messages without distortion. The project called "Distance Education for Women's Development & Empowerment" aimed to address some of these problems. In order to enhance the capacity of field level functionaries and other development related practitioners, the Department of Women and Child Development initiated a collaborative project with Indira Gandhi National Open University (IGNOU) and the Indian Space Research Organisation (ISRO) to develop a certificate course, using distance education, on Women's Group Mobilisation and Empowerment.

Δ Support to training and employment programme for women (STEP)

The STEP programme was launched in 1987 to provide skills and new knowledge to poor and asset less women working in the traditional sectors, such as agriculture, animal husbandry, dairying, fisheries, handlooms, handicrafts, khadi and village industries, sericulture, social forestry and wasteland development to enhance and broaden their employment opportunities including self-employment and development of entrepreneurial skills. Women beneficiaries were organised into groups or co-operatives. A package of services, such as, extension, inputs, market linkages, etc. was provided as well as linkages with credit for transfer of assets. The project showed that women in the dairying sector have been receiving the most support, followed by handlooms sector, handicrafts, sericulture and poultry.

Δ Establishment of employment-cum-income generation-cum-production units

The employment-cum-income generation-cum-production units programme was launched in 1982 with funding from NORAD (Norwegian Agency for Development Cooperation). Its aim was to improve the lives of poor women through providing financial assistance to women's development corporations, public sector corporations, autonomous bodies and voluntary organisations so that they could train rural women, mostly in trades that were non-traditional to them, and to ensure their employment in these areas. The trades included electronics, watch assembling, computer programming, garment making, secretarial practices, community health work, embroidery and weaving. Since the start of the project, 1477 training projects benefiting 2.28 lakh women have been approved.

Δ Rashtriya Mahila Kosh

The Rashtriya Mahila Kosh (RMK) is a registered society sponsored by the Department of Women and Child Development, Ministry of Human Resources Development, Government of India. It was established in 1993. The main objective of RMK is to facilitate credit support or microfinance to poor women, as an instrument of socio-economic change and development. RMK mainly channels its support through non-government organisations, women development corporations, cooperative societies and indira mahila block samities (under the Indira Mahila Yojana) and suitable state government agencies. RMK encourages the formation and promotion of women's SHGs by providing interest-free loans, particularly in relatively un-served areas. Rs.4000 is made available for the formation of one SHG, with up to Rs.1 lakh being made available to NGOs for the formation and stabilisation of 25 SHGs over a period of one year. During 2000-2001, the RMK has sanctioned Rs.50.40 lakh to 67 organisations under the scheme.

Δ Others

NABARD has constituted a "Credit and Financial Service Fund" to support credit innovations to improve the outreach of credit to the rural poor. To date, over 3,70,490 rural poor have been linked to commercial banks, regional rural banks and co-operative banks in 22 states and 2 union territories across India (Ramachandran, 2002).

A number of Non Governmental Organisations (NGOs) in India have implemented programmes for empowerment of women, with a special focus on income generation through organising women SHGs (thrift and credit), linking members to institutional credit and training programmes, such as, in crop production, value addition and marketing. NGOs such as BAIF, MYRADA and SEWA have several years of experience in implementing such programmes and are also training staff of other organisations in skills related to implementing development programmes for women.

While most of the capacity building programmes implemented by the government and NGOs could be classified as gender neutral or gender ameliorative, some NGOs have taken up more substantial programmes that could be called as gender transformative training of poor women. Examples of this include, training programmes of NGOs such as Action India, Jagori and Nirantar in New Delhi, Deccan Development Society and ASMITA in Andhra Pradesh, RUSEC in Tamil Nadu, Prayas, Astah, SARTHI and PEDO in Rajstahan and Gujarat and SPARC in Maharashtra.

Δ National Perspective Plan for Women 1988-2000 A.D.

To boost up the programmes for women's development, a National Perspective Plan for Women (1988-2000 A.D.) was brought out by the Department of Women and Child Development, Ministry of Human Resource Development.

The plan pays special attention to the rural women who suffer from double discrimination. The plan does not seek more investment or more resources but gives a new thrust and responsiveness to developmental programmes at all levels.

The National Perspective Plan's main aim is to promote holistic perspective to the development of women. Some of the main recommendations of the National Perspective Plan are as follows:

1. While programme for women will continue to be implemented by different ministries, there is need for a strong interministerial co-ordination and monitoring body in the Department of Women and Child Development.

2. Education to girls should be given priority and awareness needs to be generated regarding the necessity of educating girls so as to prepare them to contribute effectively to the socio-economic development of the country.

3. There is strong need to eliminate all forms of discrimination in employment especially to eliminate wage differentials between men and women.

4. The Planning Commission and all ministries and government departments must have a women's cell.

5. In order to change the attitudes towards women and girls and to raise the social consciousness of the country, a conscious strategic change is required in national media and communication effort.

6. Law drafting technologies and enforcement mechanism including police, judiciary and other components need to be reviewed, sensitised and strengthened so as to provide equality and justice.

7. Government should effectively secure participation of women in decision-making process at National, State and Local levels. This would imply use of special measures for recruitment of women candidates.

8. 30% reservation should be provided at Panchayat and at district level for women.

9. There is urgent need to improve the effectiveness of voluntary action.

Δ The National Policy for Empowerment of Women

The Government of India has declared 2001 as Women's Empowerment year. The national policy of empowerment of women has set certain clear-cut goals and objectives. The policy aims at upliftment, development and empowerment in socio-economic and politico-cultural aspects, by creating in them awareness on various issues in relation to their empowerment.

The following are the specific objectives of National Policies particularly of rural folk on Empowerment of women in India.

i. Creating an environment through positive economic and social policies for full development of women to enable them to realize their full potential.

ii. The de-jure and de-facto enjoyments of all human rights and fundamental freedom by women on equal basis with men in all political, economic, social, cultural and civil spheres.

iii. Equal access to participation and decision making of women in social political and economic life of the nation.

iv. Equal access to women to health care, quality education at all levels, career and vocational guidance, employment, equal remuneration, occupational health and safety, social security and public life etc.,

v. Strengthening legal systems aimed at elimination of all forms of discrimination against women.

vi. Changing societal attitudes and community practices by active participation and involvement of both men and women.

vii. Ministering a gender perspective in the development process.

viii. Elimination of discrimination and all forms of violence against women and the girl child.

ix. Building and strengthening partnerships with civil society, particularly women's organizations.

Δ Women Empowerment Schemes

1. Beti Bachao Beti Padhao Scheme
2. One Stop Centre Scheme
3. Women Helpline Scheme
4. UJJAWALA: A Comprehensive Scheme for Prevention of trafficking and Rescue, Rehabilitation and Re-integration of Victims of Trafficking and Commercial Sexual Exploitation
5. Working Women Hostel
6. Rajiv Gandhi National Creche Scheme For the Children of Working Mothers
7. SWADHAR Greh (A Scheme for Women in Difficult Circumstances)
8. IGMSY scheme
9. Support to Training and Employment Programme for Women (STEP)
10. NARI SHAKTI PURASKAR
11. Mahila police Volunteers
12. Mahila E-Haat

1. Beti Bachao Beti Padhao Scheme:

- It was Launched on 22nd January 2015
- Main aim-To generate awareness of welfare service meant for girl child and women.
- The Overall Goal of the Beti Bachao, Beti Padhao(BBBP) Scheme is to Celebrate the Girl Child & Enable her Education. The objectives of the Scheme are as under:-
- Prevent gender biased sex selective elimination
- Ensure survival & protection of the girl child
- Ensure education of the girl child

The Beti Bachao Beti Padhao (BBBP) initiative has two major components. i) Mass Communication Campaign and ii) Multi-sectoral action in 100 selected districts (as a pilot) with adverse CSR, covering all States and UTs.

(a) Mass Communication Campaign on Beti Bachao Beti Padhao

The campaign aims at ensuring girls are born, nurtured and educated without discrimination to become empowered citizens of this country. The Campaign interlinks National, State and District level interventions with community level action in 100 districts, bringing together different stakeholders for accelerated impact.

(b) Multi-Sectoral interventions in 100 Gender Critical Districts covering all States/UTs:-

Coordinated & convergent efforts are undertaken in close coordination with MoHFW and MoHRD to ensure survival, protection and education of the girl child. The District Collectors/Deputy Commissioners (DCs) lead and coordinate actions of all departments for implementation of BBBP at the District level. Mulit-sectoral interventions includes:

i) Ministry of WCD: Promote registration of pregnancies in first trimester in Anganwadi Centres (AWCs); Undertake training of stakeholders; Community mobilization & sensitization; Involvement of gender champions; Reward & recognition of institutions & frontline workers.

ii) Ministry of Health & Family Welfare: Monitor implementation of Pre-Conception and Pre-Natal Diagnostic Techniques (PCP&DT) Act, 1994; Increased institutional deliveries; Registration of births; Strengthening PNDT Cells; Setting up Monitoring Committees

iii) Ministry of Human Resource Development: Universal enrolment of girls; Decreased drop-out rate; Girl Child friendly standards in schools; Strict implementation of Right to Education (RTE); Construction of Functional Toilets for girls.

Δ What can we all do as individuals:

- Celebrate the birth of girl child in the family and community
- Take pride in daughters and oppose the mentality of 'Bojh' and 'Paraya Dhan'.
- Find ways to promote equality between boys and girls.
- Secure admission to & retention of girl child in schools.
- Engage men and boys to challenge gender stereotypes and roles.
- Educate and sensitize our sons to respect women and girls as equal members of society.
- Report any incident of sex determination test.
- Strive to make neighborhood safe & violence-free for women & girls.
- Oppose dowry and child marriage within the family and community.
- Advocate simple weddings.
- Support women's right to own and inherit property.
- Encourage women to go out, pursue higher studies, work, do business, access public spaces freely etc.
- Mind his language and be sensitive to women and girls.

2. One Stop Centre Scheme:

- Ministry of Women and Child Development (MWCD), has formulated a Centrally Sponsored Scheme for setting up One Stop Centres (OSC), to be funded from the Nirbhaya Fund.

- These Centres will be established across the country to provide integrated support and assistance under one roof to women affected by violence, both in private and public spaces in a phased manner. In the first phase, one OSC will initially be established in each State/UT to facilitate access to an integrated range of services including medical, legal, and psychological support.

- The OSC will be integrated with 181 and other existing helplines. The scheme is being implemented through States/UTs from 1st April 2015.

- For the implementation of the Scheme, guidelines have been developed to support stakeholders/agencies who would be involved in implementing the Scheme, as well as State, district and grassroot level functionaries. The guidelines are intended to serve as a reference manual for officials at the National and State/ Union Territory (UT) levels for policy guidance and monitoring. The guidelines list the services to be provided under the Scheme, steps and processes/procedures such as monitoring mechanism at various levels, fund flow, reporting, standard operating procedures for handholding of women at OSC etc

The implementation guidelines have a prescribed proforma, for the submission of proposal for the establishment OSCs by States/UTs. The proposals of States/UTs will be examined by a Programme Approval Board (PAB) to be constituted in the Ministry of Women and Child Development.

Provision for Shelter under OSC

- Women affected by violence along with their children (girls of all ages and boys up till 8 years of age) can avail temporary shelter at OSC for a maximum period of 5 days. The admissibility of any woman to the temporary shelter would be at the discretion of Centre Administrator.

- The women accessing temporary shelter at OSC would be provided with basic facilities i.e. food, medicine, clothes etc. A basic Kit having soap, shampoo, hair oil, sanitary pads, sewing kit, comb, tooth brush, tooth paste and diapers (in case of infants) etc will be provided to every women availing shelter facility at OSC. At any given time, OSC will provide shelter facility to maximum number of 5 women. The cost of each Kit should not exceed Rs. 100. The list of items is indicative and State may adopt this as per the local requirements.

3. Women Helpline Scheme:

The scheme of Universalisation of Women Helpline is exclusively designed to support women affected by violence, both in private and public spaces, including in the family, community, workplace etc. Women who are victims of physical, sexual, emotional, psychological and economic abuse, irrespective of age, class, caste, education status, marital status, race, culture, and geography will be provided support. In addition, woman facing any kind of violence due to attempted honour related crimes, acid attacks, witch hunting, sexual harassment, child sexual abuse, trafficking etc will also be provided with immediate and emergency services. There shall be no discrimination of any kind which affects the treatment of the aggrieved. This is specifically with reference to married women/ women in consensual sexual relationships who are raped by their intimate partners, sex workers and transgenders who might be sexually assaulted but are refused treatment due to patriarchal mindsets and prejudices.

The Women Helpline (Helpline) will provide 24 hour emergency response to all women affected by violence both in public and private sphere. All the existing emergency services such as Police (100), Fire (101), women helpline (1091), hospital/Ambulance (102), Emergency Response Services (108), NALSA Helpline for Free Legal Service (15100) and Child helpline (1098) would be integrated with this women helpline. The proposed Women Helpline will utilize the infrastructure of existing Chief Minister Helpline functioning in some States through 181 as well as that of 108 services. It will be established in every State/ UT. Following are some of the significant objectives of the Women's Helpline:

- Provide toll-free 24-hours telecom service to women affected by violence seeking support and information.
- Facilitate crisis intervention through referral to police/ Hospital/ Ambulance services
- Provide information about the appropriate support services available to the woman affected by violence, in her particular situation within the local area in which she resides or is employed.
- Creation and maintenance of a comprehensive referral database by the Helpline within its local area.

Presently, the telecom access to emergency services in India such as 181 (Women in distress), 1091 (police helpline), 100 (Police), 101 (Fire), 102 (hospital/ Ambulance) & 108 (Emergency Response Services) are being provided by Telecom Service Providers (TSPs). As per the National Numbering Plan 2003, these are Category-I services i.e. these are the mandatory services that are to be provided by all the TSPs. Whereas access to 100, 101 and 102 numbers is Restricted (these are the services to be accessible at least within local area); access to 108 and 181 has been defined as Unrestricted (these are the services, which shall be accessible from anywhere, national or international).

4. Ujjawala: (Effective 1st April, 2016)
A Comprehensive Scheme for Prevention of trafficking and Rescue, Rehabilitation and Re-integration of Victims of Trafficking and Commercial Sexual Exploitation

A. Introduction:
1. Trafficking of women and children for commercial sexual exploitation is an organized crime that violates basic human rights. India has emerged as a source, destination and transit for both in-country and cross border

trafficking. The problem of trafficking of women and children for commercial sexual exploitation is especially challenging due to its myriad complexities and variation. Poverty, low status of women, lack of a protective environment etc. are some of the causes for trafficking.

2. A multi sectoral approach is needed which will undertake preventive measures to arrest trafficking especially in vulnerable areas and sections of population; and to enable rescue, rehabilitation and reintegration of the trafficked victims.

3. Keeping the above issues and gaps in mind the Ministry has formulated a Central Scheme "Comprehensive Scheme for Prevention of Trafficking for Rescue, Rehabilitation and Re-Integration of Victims of Trafficking for Commercial Sexual Exploitation-Ujjawala". The new scheme has been conceived primarily for the purpose of preventing trafficking on the one hand and rescue and rehabilitation of victims on the other.

B. Objective of the Scheme

• To prevent trafficking of women and children for commercial sexual exploitation through social mobilization and involvement of local communities, awareness generation programmes, generate public discourse through workshops/seminars and such events and any other innovative activity.

• To facilitate rescue of victims from the place of their exploitation and place them in safe custody.

• To provide rehabilitation services both immediate and long-term to the victims by providing basic amenities/needs such as shelter, food, clothing, medical treatment including counselling, legal aid and guidance and vocational training.

• To facilitate reintegration of the victims into the family and society at large

• To facilitate repatriation of cross-border victims to their country of origin.

C. Target Group/Beneficiaries

• Women and children who are vulnerable to trafficking for commercial sexual exploitation.

• Women and children who are victims of trafficking for commercial sexual exploitation.

D. Implementing Agencies

The implementing agencies can be the Social Welfare/Women and Child Welfare Department of State Government, Women's Development Corporations, Women's Development Centres, Urban Local Bodies, reputed Public/Private Trust or Voluntary Organizations. The organization must have adequate experience in the field of trafficking, social defence, dealing with women and children in need of care and protection, children in conflict with law, etc

5. Working Women Hostel: A Scheme to Provide Safe and Affordable Accommodation to Working Women (As amended June, 2015)

1. Introduction

With the progressive change in the socio-economic fabric of the country more and more women are leaving their homes in search of employment in big cities as well as urban and rural industrial clusters. One of the main difficulties faced by such women is lack of safe and conveniently located accommodation. The Government of India being concerned about the difficulties faced by such working women, introduced a scheme in 1972-73 of grant-in-aid for construction of new/ expansion of existing buildings for providing hostel facilities to working women in cities, smaller towns and also in rural areas where employment opportunities for women exist. Based on an evaluation, the existing scheme has been revised to promote availability of safe and conveniently located accommodation for working women who need to live away from their families due to professional commitments.

2. Objectives

The objective of the scheme is to promote availability of safe and conveniently located accommodation for working women, with day care facility for their children, wherever possible, in urban, semi urban, or even rural areas where employment opportunity for women exist. To achieve this objective, the scheme will assist projects for construction of new hostel buildings, expansion of existing hostel buildings and hostel buildings in rented premises. The working women's hostel projects being assisted under this scheme shall be made available to all working women without any distinction with respect to caste, religion, marital status etc., subject to norms prescribed under the scheme. While the projects assisted under this scheme are meant for working women, women under training for job may also be accommodated in such hostels subject to the condition that taken together, such trainees should not occupy more than 30% of the total capacity the hostel and they may be accommodated in the hostels only when adequate numbers of working women are not available. Children of working women, up to the age of 18 years for girls and up to the age of 5 years for boys may be accommodated in such hostel with their mothers.

3. Beneficiaries

Following categories of working women and their children will be covered under this Scheme:

(i) Working women, who may be single, widowed, divorced, separated, married but whose husband or immediate family does not reside in the same city/area. Particular preference may be given to women from disadvantaged sections of the society. There should be also provision for reservation of seats for physically challenged beneficiaries.

(ii) Women who are under training for job provided the total training period does not exceed one year. This is only on the condition that there is vacancy available after accommodating working women. The number of women under training for job should not exceed 30% of the total capacity.

(iii) Girls up to the age of 18 years and boys up to the age of 5 years, accompanying working mothers will be provided accommodation, with their mothers. Working mothers may also avail of the services of the Day Care Centre, as provided under the scheme.

6. Rajiv Gandhi National Creche Scheme For the Children of Working Mothers:

I. Introduction

The Government's sustained initiative on education and employment of women has resulted in increased opportunities for their employment, and more and more women are now in gainful employment, working within or outside their homes. The growing industrialization and urban development has led to increased migration into the cities. The past few decades have shown a rapid increase in nuclear families and breaking up of the joint family system. Thus the children of these women, who were earlier getting support from relatives and friends while their mothers were at work, are now in need of day care services which provide quality care and protection for the children. Children who used to grow up in the secure and warm laps of their grandmothers and aunts are now confronted with an insecure and neglected environment; therefore women need a safe place for their children in their absence. It has become necessary to provide support to the young children in terms of quality, substitute care and other services while the mothers are at work. Effective day care for young children is essential and a cost effective investment as it provides support to both mothers and young children. Lack of proper day-care services is, often, a deterrent for women to go out and work. Hence, there is an urgent need for improved quality and reach of day care services/crèches for working women amongst all socio-economic groups both in the organized and unorganized sectors.

II. Definition

A crèche is a facility which enables parents to leave their children while they are at work and where children are provided stimulating environment for their holistic development. Crèches are designed to provide group care to children, usually up to 6 years of age, who need care, guidance and supervision away from their home during the day.

III. Objectives

(i) To provide day-care facilities for children (6 months to 6 years) of working mothers in the community.

(ii) To improve nutrition and health status of children.

(iii) To promote physical, cognitive, social and emotional development (Holistic Development) of children.

(iv) To educate and empower parents /caregivers for better childcare.

IV. Services

The scheme will provide an integrated package of the following services:

(i) Day-care Facilities including Sleeping Facilities.

(ii) Early Stimulation for children below 3 years and Pre-school Education for 3 to 6 years old children.

(iii) Supplementary Nutrition (to be locally sourced)

(iv) Growth Monitoring.

(v) Health Check-up and Immunization.

V. Target Group

The scheme focuses on children of 6 months to 6 years, of working women in rural and urban areas who are employed for a minimum period of 15 days in a month, or six months in a year.

VI. Coverage

The Scheme has a pan India coverage. Preference would be given to poor children and children with special nutritional needs. As on January 2015, there are 23,293 functional crèches. This Scheme will continue as a Central Sector Scheme in rural and urban areas.

In the first year of implementation of the revised Scheme, the agencies will undertake an exercise to upgrade the infrastructure in the crèches to meet the requirements of the revised Scheme. In this period the agencies will also undertake intensive inspections and weed out non-functional and non-performing crèches in these areas.

VII. Number of Beneficiaries and Functionaries

Ideally the number of children in the crèche should not be more than 25. Of these, at least 40 percent of children should, preferably, be below 3 years of age.

It is important that adequate trained worker and helper are available to provide day care facilities and to supervise the functioning of the crèche. In addition to crèche worker, there should be one crèche helper looking after children.

7. SWADHAR Greh (A Scheme for Women in Difficult Circumstances), 2015

A. Introduction

Recognizing the need to prevent women from exploitation and to support their survival and rehabilitation, the scheme of Short Stay Home for women and girls was introduced as a social defense mechanism, by the then Department of Social Welfare in 1969. The scheme is meant to provide temporary accommodation, maintenance and rehabilitative services to women and girls rendered homeless due to family discord, crime, violence, mental stress, social ostracism or are being forced into prostitution and are in moral danger. Another scheme with the similar objectives namely Swadhar-A Scheme for Women in Difficult Circumstances was launched by the Department of Women and Child Development in 2001-02. The scheme through the provisions of shelter, food, clothing, counseling, training, clinical and legal aid aims to rehabilitate such women in difficult circumstance. Centre for Market Research and Social Development, New Delhi conducted an evaluation in 2007 to assess the performance of both the schemes. The evaluation report while citing the effectiveness and positive impact of measures adopted under the schemes for counseling and rehabilitation found that the profile and category of residents, admission procedure, counseling, quality of service, vocational training, rehabilitation and follow up procedure are almost similar in both the schemes. It, therefore, recommended merger of these two schemes for better functioning and outcomes with lesser administrative burdens and procedures. It also recommended that the new scheme should focus on establishing one such home in every district.

The positive findings of the evaluation study has encouraged the Ministry to propose this new scheme that would target the women victims of unfortunate circumstances who are in need of institutional support for rehabilitation so that they could lead their life with dignity.

B. **Vision**

The scheme envisions a supportive institutional framework for women victims of difficult circumstances so that they could lead their life with dignity and conviction. It envisages that shelter, food, clothing, and health as well as economic and social security are assured for such women. It also envisions that the special needs of these women are properly taken care of and under no circumstances they should be left unattended or abandoned which could lead to their exploitation and desolation.

C. **Objectives**

Under the Scheme, Swadhar Greh will be set up in every district with capacity of 30 women with the following objectives:

(a) To cater to the primary need of shelter, food, clothing, medical treatment and care of the women in distress and who are without any social and economic support.

(b) To enable them to regain their emotional strength that gets hampered due to their encounter with unfortunate circumstances.

(c) To provide them with legal aid and guidance to enable them to take steps for their readjustment in family/society.

(d) To rehabilitate them economically and emotionally.

(e) To act as a support system that understands and meets various requirements of women in distress.

(f) To enable them to start their life afresh with dignity and conviction.

For big cities and other districts having more than 40 lakh population or those districts where there is a need for additional support to the women, more than one Swadhar Greh could be established. The capacity of Swadhar Greh could be expanded up to 50 or 100 on the basis of need assessment and other important parameters.

D. Strategies

The objectives cited above would be pursued adopting the following strategies:

a) Temporary residential accommodation with the provision of food, clothing, medical facilities etc.

b) Vocational and skill up gradation trainings for economic rehabilitation of such women

c) Counseling, awareness generation and behavioral trainings

d) Legal aid and Guidance

e) Counseling through telephone

E. Beneficiaries

The benefit of the component could be availed by women above 18 years of age of the following categories:

a) Women who are deserted and are without any social and economic support;

b) Women survivors of natural disasters who have been rendered homeless and are without any social and economic support;

c) Women prisoners released from jail and are without family, social and economic support;

d) Women victims of domestic violence, family tension or discord, who are made to leave their homes without any means of subsistence and have no special protection from exploitation and/ or facing litigation on account of marital disputes; and

e) Trafficked women/girls rescued or runaway from brothels or other places where they face exploitation and Women affected by HIV/ AIDS who do not have any social or economic support. However such women/ girls should first seek assistance under UJJAWALA Scheme in areas where it is in operation.

Women affected by domestic violence could stay up to one year. For other categories of women, the maximum period of stay could be up to 3 years. The older women above the 55 years of age may be accommodated for maximum period of 5 years after which they will have to shift to old age homes or similar institutions. Swadhar Greh facilities could also be availed by the children accompanying women in the above categories. Girls up to the age of 18 years and boys up to the age of 8 years would be allowed to stay in the Swadhar Greh with their mothers. (Boys of more than 8 years of age need to be shifted to the Children Homes run under JJ Act/ICPS.)

8. Indira Gandhi Matritva Sahyog Yojana (IGMSY): A Conditional Maternity Benefit Scheme
1. Introduction:

i. The vulnerable condition of the pregnant women belonging to poor and economically deprived families across the country is well recognised. In the Eleventh Five Year Plan document (Vol.II), the Planning Commission has noted that "Poor women continue to work to earn a living for the family right upto the last days of their pregnancy, thus not being able to put on as much weight as they otherwise might. They also resume working soon after childbirth, even though their bodies might not permit it—preventing their bodies from fully recovering, and their ability to exclusively breastfeed their new born in the first six months. Therefore, there is urgent need for introducing a modest maternity benefit to partly compensate for their wage loss."

ii. Under-nutrition, especially in infant and young children, adolescent girls and women results in increased susceptibility to infections, slow recovery from illnesses, cumulative growth and development deficits leading to reduced productivity and a heightened risk of adverse pregnancy outcomes for women. A woman's nutritional status has important implications for her health as well as the health and development of her children. A woman with poor nutritional status, as indicated by a low body mass index (BMI), short stature, anaemia, or other micronutrient deficiencies, has a greater risk of obstructed labour, having a baby with a low birth weight and adverse pregnancy outcomes resulting in death due to postpartum hemorrhage, illness for herself and her baby and adversely affecting lactation.

iii. In India, high levels of under-nutrition and anaemia in adolescent girls and women are compounded by early marriage, early child bearing and inadequate spacing between births. Girls and women often face an inter-generational cycle of under nutrition compounded by multiple deprivations-gender discrimination, poverty and exclusion. This vicious cycle needs to be addressed through multisectoral interventions. Due to increased nutritional needs during pregnancy and lactation, the pregnant and lactating mothers require greater nutritional support, especially in settings where levels of under nutrition and anaemia are already high. During this period, mothers require access to health care services, enhanced food and nutrient intake, family care, skilled counselling support and a hygienic environment. Therefore, improvement in nutritional status of women especially during pregnancy and lactation, requires multi-sectoral, concerted, convergent and supportive actions.

iv. Maternal under-nutrition is a major challenge in India with more than one third (35.6%) having low Body Mass Index (BMI). Early marriage, early child bearing and frequent pregnancy adversely affect the maternal nutritional status. According to the NFHS-III, 58 per cent women were married before the legal age of 18 years and three quarters (74%) were married before reaching the age of 20 years. Around 30% women aged 25-49 years gave first birth before the age of 18 years and 50% gave first birth when they were at the age of 20 years. The prevalence of anaemia for ever-married women has increased from 52 % in NFHS-2 to 56 % in NFHS-3. According NFHS-3, 69.5 children aged 6-59 months reported any anaemia with 26.3 per cent having mild anaemia, 40.2 per cent moderate anaemia and 2.9 per cent severe anaemia. Anaemia tends to increase with the number of children ever born and decreases with education and the household's wealth. Anaemia is more prevalent for women who are breastfeeding (63 %) and women who are pregnant (59 %) than for other women (53 %). Therefore, the anaemia situation has worsened over time for both women and young children.

v. The promotion of early and exclusive breastfeeding for the first six months and appropriate complementary feeding continue to be major challenges. According to NFHS-III, only about 25 % of the babies are initiated into breastfeeding within one hour of birth. Only 46 % of children under five months of age are exclusively breastfed. It is significant that complementary feeding has increased substantially. The percentage of infants (between 6 to 9 months) receiving complementary feeding, along with breast milk, increased significantly from 33.5 % to 55.8 % during the period 1998-99 (NFHS-II) to 2005-06 (NFHSIII). According to Lancet 2003-India Analysis, 16 % of under 5 child mortality in India can be averted by ensuring universal exclusive breastfeeding for the first six months of the infant's life. Another 5 % can be reduced by promoting the universal practice of appropriate complementary feeding.

vi. Maternal mortality is defined by NFHS-III as the death of a woman during pregnancy or delivery or within 42 days of the end of pregnancy from a pregnancy-related cause. Approximately 30 million women in India are pregnant annually, and 27 million have live births. Of these, nearly 136,000 maternal deaths occur annually, most of which can be prevented. According to data from the Registrar General of India quoted in Special Bulletin in Maternal Mortality in India (April 2009), maternal mortality ratio in India is 254 as reported between 2004 and 2006. This is derived as the proportion of maternal deaths per 100,000 live births reported under SRS.

vii. Infant and Child Mortality: Around 1.7 million children in India do not reach their first birthday, of these 1.2 million die within the first month.

According to the NFHS-3, infant mortality in India has declined from 77 deaths per 1,000 live births in 1991-95 (10-14 years before the survey) to 57 deaths per 1,000 live births in 2001-05 (0-4 years before the survey), thus implying an average rate of decline of 2 infant deaths per 1,000 live births per year. All other measures of infant and child mortality also show declining trends during the years before the survey. By comparing the estimates for the period 10-14 years before the survey with the estimates for the period 0-4 years before the survey, it is seen that the neonatal mortality rate has decreased by 12 deaths per 1,000 live births (from 51 to 39), the postneonatal mortality rate has decreased by 7 deaths per 1,000 live births (from 25 to 18), and the child mortality rate (at age 1-4 years) has decreased by 14 deaths per 1,000 children age 1 (from 32 to 18).

viii. Antenatal care: Among mothers who gave birth in the five years preceding the NFHS-III survey, almost three-quarters received antenatal care from a health professional (50% from a doctor and 24% from other health personnel). Younger women were more likely than older women to receive antenatal care, as were women with more education and women having their first child. Less than half of women received antenatal care during the first trimester of pregnancy, as is recommended. Another 22 % had their first visit during the fourth or fifth month of pregnancy. Just over half of mothers had three or more antenatal care visits; urban women were much more likely to receive three or more visits than women in rural areas. For 65 % of births, mothers received iron and folic acid supplements, but only 23 % consumed them for the recommended 90 days or more. Three in four mothers received two or more doses of tetanus toxoid vaccine.

ix. Immunisation of children: As per National Family Health Survey (NFHS-3) Less than half (44 %) of children 12-23 months are fully vaccinated against the six major childhood illnesses: tuberculosis, diphtheria, pertussis, tetanus, polio, and measles. However, most children are at least partially vaccinated: only 5 % have received no vaccinations at all. 78 % of children have received a BCG vaccination, and the same have received at least the recommended three doses of polio vaccine. However, only 59 % have been vaccinated against measles, and only 55 % have received all the recommended doses of DPT.

x. In view of the above situations, there is an emergent need to address the nutritional deficits of the pregnant and lactating mother. This could be promoted by providing maternal support, counseling and services in an enabling environment with a view to enhance the demand and utilization of existing maternal and child care services. Such a support could also be through a direct cash transfer system on achieving certain conditionality's which could be used by the beneficiary for her own care and that of her child.

Δ **"Indira Gandhi Matritva Sahyog Yojana (IGMSY)"-Conditional Maternity Benefit(CMB) Scheme would be implemented through**

The platform of Integrated Child Development Services(ICDS) Scheme. The focal point of implementation would be the Anganwadi Centre(AWC) at the village.

4. Objectives

To improve the health and nutrition status of pregnant & lactating women and infants by:

- Promoting appropriate practices, care and service utilisation during pregnancy, safe delivery and lactation;
- Encouraging the women to follow (optimal) IYCF practices including early and exclusive breast feeding for the first six months;
- Contributing to better enabling environment by providing cash incentives for improved health and nutrition to pregnant and lactating mothers.

5. Target Group

Pregnant Women of 19 years of age and above for first two live births (benefit for still births would be as per the guidelines of scheme) All Government/PSUs (Central & State) employees would be excluded from the scheme as they are entitled for paid maternity leave.

9. Support to Training and Employment Programme for Women (STEP)

The Ministry has been administering 'Support to Training and Employment Programme for Women (STEP) Scheme' since 1986-87 as a 'Central Sector Scheme'. The STEP Scheme aims to provide skills that give employability to women and to provide competencies and skill that enable women to become self-employed/entrepreneurs. The Scheme is intended to benefit women who are in the age group of 16 years and above across the country. The grant under the Scheme is given to an institution/ organisation including NGOs directly and not the States/ UTs. The assistance under STEP Scheme will be available in any sector for imparting skills related to employability and entrepreneurship, including but not limited to the Agriculture, Horticulture, Food Processing, Handlooms, Tailoring, Stitching, Embroidery, Zari etc, Handicrafts, Computer & IT enable services along with soft skills and skills for the work place such as spoken English, Gems & Jewellery, Travel & Tourism, Hospitality.

Revised Guidelines-2016

1. Introduction and Need for the Scheme

Government of India has set an ambitious plan training 500 million individuals by 2022 which translates to training 42 million a year. For this purpose, India's vocational training infrastructure needs to be expanded to meet the diverse and many skill requirements of the population. There has been recent concern about the decline in women's workforce participation in India. Simultaneously, women have become more aspirational and are ready to contribute equally to the economy.

MWCD, through its STEP Programme, has been addressing special situation of poor women and women in remote areas who are not in a position to move out of their immediate surroundings and go to a formal skill centre to acquire training. The STEP Programme a 100% Central Sector Scheme is under implementation since 1986-87. Training is provided to poor and marginalized women in traditional trades which are largely in the informal sector. The programme strives to build upon the traditional knowledge of women and convert it into sustainable livelihood capacitation.

STEP Guidelines 2014 are revised on the basis of learning of the rigorous process of scrutiny of thousands of proposals that were received in response to the 2014 Guidelines as well as the Gazette notification issued by the Ministry of Skill Development & Entrepreneurship and NITI Aayog's Guidelines for implementation of Centrally Sponsored Schemes/Central Sector Schemes through NGOs. A separate section has been added as guidance to the applicant organizations.

2. Objectives of the Scheme

The scheme has 2 fold objectives viz.

a. To provide skills that give employability to women.

b. To provide competencies and skills that enable women to become self-employed/entrepreneurs.

3. Target Groups

The scheme is intended to benefit women who are in the age group of 16 years and above.

4. Eligible Organizations/Project Implementing Agencies (PIAS)

Grants-in-aid under the STEP programme may be given to an institution having a distinct legal entity as under:

(a) Institutions or organizations set up as Autonomous Organization under a specific statute or as a Society registered under the Societies Registration Act, 1860 or Indian Trusts Act, 1882 (Not for profit) or other statutes.

(b) Voluntary Organizations or Non-Government Organizations registered under the Societies Registration Act, Indian Trust Act carrying out activities which promote the objectives of the STEP programme, with adequate financial and other resources, credibility and experience of the type of activities to be undertaken.

(c) Co-operative Societies.

5. Criteria for Selection of Project Implementing Agencies (PIAS)

(a) The NGO should have signed up in the NGO-Partnership (NGO-PS) Portal of the NITI Aayog with all self-declared details and should have obtained a Unique ID. The Unique ID should be mandatorily quoted by the NGO at the time of application for grants. NGOs shall update their data base in the Portal every year.

(b) The NGO should also have signed up in the portal of this Ministry with the Unique ID of NITI Aayog.

(c) The organization must be in existence at least for 3 years and have carried out activities for imparting skills related to employability and entrepreneurship during last 3 years. At the time of filing the application, the organization must have a positive net worth in at least 2 previous years.

6. Trades Covered

(a) Assistance under the STEP Scheme will be available in any sector for imparting skills related to employability and entrepreneurship as identified by the Ministry of Skill Development & Entrepreneurship (MSDE) including but not limited to the following:

- Agriculture
- Horticulture
- Food Processing
- Handlooms
- Traditional crafts like Embroidery, Zari etc.
- Handicrafts
- Gems & Jewellery
- Travel & Tourism, Hospitality

(b) Soft skills (which would include computer literacy, language and workplace inter-personal skills relevant for the sector/trade) would be an integral part of the skills training process and must be suitably integrated into the course modules.

(c) All Skill Development courses offered under the scheme framework must conform to the National Skill Qualification Framework (NSQF). Government funding would not be available for any training or educational programme/course if it is not NSQF compliant.

7. Requirements/Input Standards for PIAS

(a) Input Standards

The following inputs are essential to be followed so as to ensure that adequate training infrastructure and capacity exists:

i) The overall training infrastructure specially the training aids and equipment being as per industry benchmarks.

ii) Trainers with suitable qualifications/experience being hired and should have undergone Training of Trainers (ToT).

iii) Industry relevant content appropriate to the learning groups and conforming to the requirements of NSQF/SDIS, being used

iv) The student and trainer enrollment linked to Aadhar.

v) Assessments being video recorded if required.

(b) Other Requirements

(i) The implementing agency will issue Photo Identity Card to all beneficiaries for the project and a copy of the same will be forwarded to the Ministry for records.

(ii) The implementing agency will obtain the bank account details (Bank Name, Branch, Account No., IFSC code, MICR code,) of all the beneficiaries and forward the same to the Ministry.

(iii) The implementing agency will obtain the photocopy of Aadhar Card of all the beneficiaries and forward the same to the Ministry.

(iv) The implementing agency would also ensure the employment/self-entrepreneurship for passed out trained candidates in accordance with the extant Guidelines of MSDE.

8. Number of Beneficiaries

Approval would be given only for such number as are considered viable and the maximum number of beneficiaries in a project shall not exceed 200.

Training will be imparted in batches with number that are manageable, within the infrastructure and training capacity of the organization and in groups that can be imparted due attention by the instructors. For trades that use equipment/ small machines/ computers, the batches will typically be determined by the number of such equipment that is available.

9. Course Duration & Training Hours

Skill development will be domain specific demand led skill training activity leading to employment or any outcome oriented activity that enables a participant to acquire a skill duly assessed and certified by an independent third party agency, and which enables her to get wage/self-employment leading to increased earnings, and/ or improved working conditions, such as getting formal certification for hitherto informal skills. The project in any case must ensure a training input spread over 200 hours in a 3 month programme and 400 hours in a 6 months programme (including practical and/or on the job training).

10. Nari Shakti Puraskar
Introduction
Every Year, Ministry of Women & Child Development celebrates International Women Day on 8th March. The significance of the International Women's Day lies in our re-affirmation of improve the condition of women, especially those at the margins of our society and empower them to take rightful place in society.

Ministry of Women and Child Development, has revised the guidelines for Women Awards for conferring on eminent women, organisations and institutions. These awards will be called "Nari Shakti Puruskars". Now, from the year 2016, 20 Nari Shakti Puruskars shall be conferred every year. The awards will be conferred on 8th March on the occasion of International Women's Day (IWD). The Award in each category shall carry a Certificate and a cash amount. The award would be given to eminent or outstanding Institutions or organizations and individuals.

With the institution of these awards all the earlier women awards given by the Ministry of Women & Child Development ceased to exist.

Objectives:
1. In the last decade, there has been concerted effort by the Government to recognise and encourage women as reflected through a National Policy for Empowerment of Women in 2001. The issues related to women has gained utmost importance and focussed attention. "Nari Shakti Puruskars" shall showcase the Government's commitment towards women with the aim of strengthening their legitimate place in the society. It will also provide an opportunity to the current generation to understand the contribution of women in building of society and the nation.
2. "Nari Shakti Puruskars" would be conferred on eminent women and institutions rendering distinguished service to the cause of women especially belonging to the vulnerable and marginalized sections of the society. The recipients would be drawn from institutions and individuals. The Ministry of Women & Child Development (herein after referred to as " the Ministry) would invite nominations from the State Governments, Union Territory Administrations, concerned Central Ministries, Non-Governmental Organizations, Universities, Institutions, private and public sector undertakings (PSUs) working for empowerment of women.

Description
- There are 20 Nari Shakti Puruskars shall be conferred every year. The recipients of the Puruskars shall be declared every year on 20th February and awards will be conferred on 8th March on the occasion of International Women's Day (IWD). The Award in each category shall carry a Certificate

and a cash amount. The award would be given to eminent or outstanding Institutions or organisations and individuals from any part of the country.

A Screening Committee and a Selection Committee would be constituted for the purpose which shall consider the achievement of the organizations and individuals nominated or recommended for the Puruskars by the prescribed authority. The outstanding contributions in the field shall be of primary consideration in identifying the recipients of Puruskars.

Eligibility Criteria for Nomination

1. The Puruskars are open to all Indian Institutions, organisations and individuals without any distinction or discrimination on ground of race, caste or creed.

2. In case of Individual category, the awardee must be above 30 years of age on 1st January of the year for which the award is to be given. The applicant should have worked in the relevant field for at least last 5 years. She should not have been a recipient earlier of the award (including Stree Shakti Puruskars).

3. The eligibility criteria for each of the category of Awards shall be as follows:

   1. Rani Rudramma Devi Award for Best Panchayat /Village Community

   2. Mata Jijabai Award for Best Urban Local body for providing services and facilities to women

   3. Kannagi Devi Award for Best State which has appreciably improved Child Sex Ratio (CSR)

   4. Rani Gaidinliu Zeliang Award for Best Civil Society Organization (CSO) doing outstanding work for the welfare and well-being of women

   5. Devi Ahilyabai Holkar Award for Best Private Sector Organization/ Public Sector Undertaking in promoting the well-being and welfare of women

   6. Rani Lakshmibai Award for Best Institution for Research & Development in the field of women empowerment

   7. Award for Courage & Bravery-2 Nos.

   8. Awards for making outstanding contributions to women's endeavour/ community work/making a difference/women empowerment-7 Nos.

11. Mahila police Volunteers
Introduction

Gender-Based Violence (GBV) faced by women both in public and private spaces, including domestic violence, sexual assault, rape, voyeurism, stalking etc are a major threat to women equality and empowerment. A gender responsive police service requires specific training, increased presence of female personnel within the police force and community outreach to integrate gender issues into policies, protocols and operational procedures.

In recent years, there has been enactment of various legislations by the Parliament which address the issue of GBV i.e. the Criminal Law Amendment Act, 2013, the Sexual Harassment of Women at Workplace (Prevention, Prohibition and Redressal) Act, 2013, the Protection of Women From Domestic Violence Act, 2005 and provided an opportunity to women facing violence to take the recourse to law. According to the latest National Crime Records Bureau data, during the year 2014, 3,37,922 incidences of crime against women (both under Indian Penal Code and other laws) were reported as against the 3,09,546 cases reported during 2013. Role of Police is pivotal in safety and security of citizens in general and women in particular. To increase the visibility of women in the police force, Home Ministry has carried forward the initiative to give 33% reservation to women in police force by implementing it in UTs and propagating in the States. There has been an increasing emphasis on gender sensitivity of police force through training programmes, performance appraisal, women police stations to tackle crime against women. A recent advisory dated 12th May, 2015 by the Home Ministry stresses on the need for sensitivity in handling women's issues.

However, it is a matter of common knowledge that women who are victim of violence or harassment may not find it easty to approach the police or other authorities for getting help or support. It would, therefore, be desirable to provide them an effective alternative for getting help and support.

In order to promote these objectives and increase focused community outreach, Government of India envisages engagement/ nomination of Mahila Police Volunteers (MPVs) in all States and UTs who will act as a link between police and community and facilitate women in distress. This will be done in a phased manner.

Vision

MPVs are envisaged as empowered, responsible, socially aware women for fostering leadership in local settings to facilitate police outreach on gender concerns. They will be an interface between the society and police.

Objectives

An MPV will serve as a public-police interface in order to fight crime against women. The broad mandate of MPVs is to report incidences of violence against women such as domestic violence, child marriage, dowry harassment and violence faced by women in public spaces. She will act as a role model for the community. An MPV is an honorary position.

Coverage

The Scheme in the first phase will be implemented on pilot basis in all States and UTs. In the first phase, two districts from every state and one district from every UT shall be chosen on the basis of the following criteria:-3

  i.   Child Sex Ratio (CSR)
  ii. Crime against Women

The final selection of the district/districts would be the responsibility of the concerned State/UT.

12. Mahila E-Haat

The Ministry of Women & Child Development launched "Mahila E-Haat" a bilingual portal on 7th March, 2016. This is a unique direct online marketing platform leveraging technology for supporting women entrepreneurs/SHGs/NGOs for showcasing the products / services which are made/manufactured/undertaken by them. It is an initiative for meeting aspirations and needs of women. This was done keeping in mind that technology is a critical component for business efficiency and to make it available to the majority of Indian women entrepreneurs / SHGs / NGOs.

Mahila E-Haat is an online marketing platform for meeting aspirations and needs of women entrepreneurs/SHGs/NGOs by showcasing their products and services. The USP of this online marketing platform is facilitating direct contact between the vendor and buyer. It is easy to access as the entire business of E-haat can be handled through a mobile.

Vision: to empower & strengthen financial inclusion of Women Entrepreneurs in the economy by providing continued sustenance and support to their creativity.

Mission: to act as a catalyst by providing a web based marketing platform to the women entrepreneurs to directly sell to the buyers.

Goal: to support 'Make in India' through digital marketing platform

The USP of this online marketing platform are:

- Facilitating direct contact between the vendors and buyers, as by displaying their contact number, address as also the basic cost of products/ services.
- Aadhar number is to be filled in mandatorily in the join us form, as a measure of id and for the payments to be received.
- For understanding how cash-less / digital transaction can be undertaken "Step by step instructions for various modes of payment: UPI, Wallets, PoS, and SMS banking (USSD)" is uploaded along with BHIM App information on the portal.
- Women need to be majorly involved in the value chain and have to be 18 years of age to display their products/ services. No illegal or contraband goods can be displayed. The vendors allowed to price their products and charge accordingly from the buyers.
- Being web based, it has unlimited reach and it is also very simple to join Mahila E-haat i.e. through the portal itself as the entire business of E-haat can be handled through a mobile. The vendor can be approached by the buyer physically, telephonically, by email, etc.

- Since its launch over 17 lakhs visitors / hits have been received by the Mahila Ehaat portal. As on date Women entrepreneurs/SHGs/NGOs from 24 states are showcasing over 2000 products/services across 18 categories viz., Clothing (Men, Women & Children), Bags, Fashion Accessories /Jewellery, Decorative and gift items, Home Décor, Carpets / Rugs /Foot mats, Baskets, Linen/ Cushion Covers, Boxes, Pottery, Grocery & Staples / Organic, Natural Products, File Folders, Industrial Products, Educational Aids, Soft Toys, Miscellaneous. This is impacting 3.50 lakh beneficiaries directly and indirectly and over 26000 SHGs. The portal is continuously being upgraded. The URL is http://mahilaehaat-rmk.gov.in.

- Mahila E-haat received the SKOCH GOLD Award on 09th Sep, 2016 and was also awarded 'SKOCH Order-of-Merit' Award, as it was adjudged as one of the "Top 100 Projects in India" for the year 2016.

- To increase visibility major PSUs, IRCTC, Nationalised Banks like SBI have given a link to Mahila E-haat on their websites.

- Information on how logistics can be handled as shared by India Post and Payment Processes and information shared by SBI is placed under useful links on the portal.

- Sensitization, advocacy, training, packing and soft intervention workshops on Mahila E-haat are organized periodically with the support of State Governments & Women Development Corporations-New Delhi, Indore, Kochi, Bangalore, Nagaland, Chandigarh, Chhattisgarh, Mumbai, Raipur, Hyderabad, Varanasi etc.

This exclusive portal is the first in the country to provide a special, focused marketing platform for women. Being a bilingual portal, it aims at financial inclusion and economic empowerment of women. This unique e-platform showcases products and services.

Δ   Child Protection & Welfare Schemes

1. Integrated Child Protection Scheme (ICPS)
2. Rashtriya Bal Kosh (National Childrens Fund)
3. Rajiv Gandhi Scheme for Empowerment of Adolescent Girls (RGSEAG) Sabla
4. Kishori Shakti Yojana

1.   Integrated Child Protection Scheme (ICPS)

Integrated Child Protection Scheme is a comprehensive scheme introduced in 2009-10 by the Government of India to bring several existing child protection programmes under one umbrella, with improved norms. The Scheme incorporates other essential interventions, which aim to address issues which were, so far, not covered by earlier Schemes. It is based on the cardinal principles of "protection of child rights" and the "best interest of the child".

The Integrated Child Protection Scheme (ICPS) is a centrally sponsored scheme aimed at building a protective environment for children in difficult circumstances, as well as other vulnerable children, through Government-Civil Society Partnership

- ## Objectives

ICPS brings together multiple existing child protection schemes of the Ministry under one comprehensive umbrella, and integrates additional interventions for protecting children and preventing harm. ICPS, therefore, would institutionalize essential services and strengthen structures, enhance capacities at all levels, create database and knowledge base for child protection services, strengthen child protection at family and community level, ensure appropriate inter-sectoral response at all levels. The scheme would set up a child protection data management system to formulate and implement effective intervention strategies and monitor their outcomes. Regular evaluation of the programmes and structures would be conducted and course correction would be undertaken.

2. **Rashtriya Bal Kosh (National Childrens Fund)**

The National Children's Fund (NCF) was set up by the Government of India during the year 1979. The National Policy for Children, announced in the year 1974 recognises children of the Nation as a 'Supremely important asset'. Their nurture and development are the responsibility of the nation. In order to help and promote various welfare and development programmes for children, the Government of India, in the International Year of the Child (1979), created National Children's Fund (NCF) with a corpus fund of rupees one lakh under the Charitable Endowment Act 1890. The Minister of State, Ministry of Women and Child Development is the Chairperson of Board of Management of NCF. The Secretary, Ministry of Women and Child Development is its Working Chairperson.

The objectives of NCF are to raise funds from individuals, institutions, Corporates and others, to promote and fund the various programmes for children who are affected by natural calamities, disasters, distress and in difficult circumstances through voluntary agencies and State Governments in unserved and underserved areas including tribal and remote areas (as per the National Charter for Children, 2003) and children of prisoners, children affected by riots, aggression, trafficking and children of prostitutes, and to implement various programmes. The secretariat of the fund is located in National Institute of Public Cooperation and Child Development (NIPCCD), an autonomous body under the aegis of the Ministry of Women and Child Development.

3. Rajiv Gandhi Scheme for Empowerment of Adolescent Girls (RGSEAG) (Sabla)

Introduction:

1. The term "Adolescence" literally means "to emerge" or "achieve identity". Its origin is from a Latin word "Adolescere" meaning, "to grow, to mature". It is a significant phase of transition from childhood to adulthood. A universally accepted definition of the concept of adolescence has not been established, but WHO has defined it in terms of age spanning between 10 to19 years. In India, the legal age of marriage is 18 years for girls and 21 years for boys. There is a high correlation between the age at marriage, fertility management and family health with education. Having regard to this and other considerations, for the purpose of this scheme, the girls in the age group between 11 to 18 years will be considered in the category of adolescent girls.

2. In India, adolescents girls (11-18 years) constitute nearly 16.75 % (Registrar General and Census Commissioner, India, 2001) of the total female population of 49.6514 crores which is approx. 8.3 crores. The female literacy rates are only 53.87% and nearly 2.74 crore girls are undernourished (33% of 8.3 crores). About 56.2% women (age 15-49), are anaemic as reflected in NFHS-3 survey. Thus, they have considerable unmet needs in terms of education, health (mainly reproductive health) and nutrition. This is largely due to the lack of targeted health services for adolescents and widespread gender discrimination that prevail and limit their access to health services as well as the practice of early marriage and child-bearing that persists and puts adolescent girls and their children at increased risk of adverse outcomes. The Constitution of India enshrines the principle of gender equality to enable the State to adopt positive measures to prevent discrimination against girl children, adolescent girls and women.

3. Adolescence is a significant period for mental, emotional and psychological development. Adolescence represents a window of opportunity to prepare for healthy adult life. During this period, nutritional problems originating earlier in life can be partially corrected, in addition to addressing the current ones. It is also the period to shape and consolidate healthy eating and life style behaviors, thereby preventing the onset of nutrition related chronic diseases in womanhood and prevalence of malnutrition in future generation. Iron deficiency anaemia is the most widespread micronutrient deficiency affecting the vulnerable groups including adolescent girls which reduces the capacity to learn and work, resulting in lower productivity and limiting economic and social development. Anaemia during pregnancy leads to high maternal and neonatal mortality and low birth weight etc. Addressing the health needs of Adolescent Girls will not only lead to a healthier and more productive women force but will also help to break the intergenerational cycle of malnutrition.

4. Within the Human Rights framework established and accepted by the global community, the rights particularly relevant to adolescents include

gender equality, right to education and health (including reproductive and sexual health) and information and services appropriate to their age, capacities and circumstances. Definite measures should to be taken to ensure these rights and also make the girls aware of their duties and responsibilities. The Adolescent Girls (AGs) need to be looked at not just in terms of their own needs as AGs but also as individuals who can be productive members of the society.

5. The Ministry of Women and Child Development, Government of India, in the year 2000 came up with scheme called "Kishori Shakti Yojna"(KSY) using the infrastructure of Integrated Child Development Services(ICDS). The objectives of the Scheme were to improve the nutritional and health status of girls in the age group of 11-18 years as well as to equip them to improve and upgrade their home-based and vocational skills; and to promote their overall development including awareness about their health, personal hygiene, nutrition, family welfare and management. The scheme provided for Rs.1.1 lakh per project per annum. 2-3 AGs per AWC are targeted under this scheme who are also provided supplementary nutrition by the state governments.

Thereafter, Nutrition Programme for Adolescent Girls (NPAG) was initiated as a pilot project in the year 2002-03 in 51 identified districts across the country to address the problem of under-nutrition among adolescent girls. Under the programme, 6 kg of free food grains per beneficiary per month are given to underweight adolescent girls.

The above two schemes have influenced the lives of AGs to some extent, but have not shown the desired impact. Moreover, the above two schemes had limited financial assistance and coverage besides having similar interventions and catered to more or less the same target groups. A need has therefore, emerged to formulate a new comprehensive scheme with richer content, merging the erstwhile two schemes that would address the multi-dimensional problems of AGs. This Scheme shall be called Rajiv Gandhi Scheme for Empowerment of Adolescent Girls (RGSEAG)--"SABLA". It would replace KSY and NPAG in the 200 selected districts. KSY would be continued (where operational) in remaining districts.

6. Rajiv Gandhi Scheme for Empowerment of Adolescent Girls-SABLA- would be implemented using the platform of ICDS Scheme through Anganwadi Centers (AWCs).

7. **Objectives**
   The objectives of the Scheme are
   i. Enable the AGs for self-development and empowerment
   ii. Improve their nutrition and health status.
   iii. Promote awareness about health, hygiene, nutrition, Adolescent Reproductive and Sexual Health (ARSH) and family and child care.

iv. Upgrade their home-based skills, life skills and tie up with National Skill Development Program (NSDP) for vocational skills

v. Mainstream out of school AGs into formal/non formal education

vi. Provide information/guidance about existing public services such as PHC, CHC, Post Office, Bank, Police Station, etc.

8. Target Group

The Scheme would cover adolescent girls in the age group of 11-18 years under all ICDS projects in selected 200 districts in all the States/UTs in the country. In order to give appropriate attention, the target group would be subdivided into two categories, viz. 11-15 & 15-18 years and interventions planned accordingly. The Scheme focuses on all out-of-school adolescent girls who would assemble at the Anganwadi Centre as per the time table and frequency decided by the States/UTs. The others, i.e., the school going girls would meet at the AWC at least twice a month and more frequently during vacations/holidays, where they will receive life skill education, nutrition & health education, awareness about other socio-legal issues etc. This will give an opportunity for mixed group interaction between in-school and out-of-school girls, motivating the latter to join school.

9. Services

An integrated package of services is to be provided to AGs that would be as follows.

Nutrition provision

ii. Iron and Folic Acid (IFA) supplementation

iii. Health check-up and Referral services

iv. Nutrition & Health Education (NHE)

v. Counseling/Guidance on family welfare, ARSH, child care practices and home management

vi. Life Skill Education and accessing public services

vii. Vocational training for girls aged 16 and above under National Skill Development Program (NSDP)

4. Kishori Shakti Yojana (KSY)
Introduction

1. Adolescence is a crucial phase in the life of woman. At this stage, she stands at the threshold of adulthood. This stage is intermediary between childhood and womanhood and it is the most eventful for mental, emotional and psychological well being. The life-cycle approach for holistic child development remains unaddressed if adolescent girls are excluded from the developmental programmes aimed at human resource development.

2. For the first time in India, a special intervention was devised for adolescent girls using the ICDS infrastructure. ICDS with its opportunity for childhood development, seeks to reduce both socio-economic and gender inequities. The Adolescent Girls (AG) Scheme under ICDS primarily aimed at breaking the inter-generational life-cycle of nutritional and gender disadvantage and providing a supportive environment for self-development.

The objectives of the Scheme are as follows:-

i)   to improve the nutritional and health status of girls in the age group of 11-18 years;

ii) to provide the required literacy and numeracy skills through the non-formal stream of education, to stimulate a desire for more social exposure and knowledge and to help them improve their decision making capabilities;

iii) to train and equip the adolescent girls to improve/ upgrade home-based and vocational skills;

iv) to promote awareness of health, hygiene, nutrition and family welfare, home management and child care, and to take all measure as to facilitate their marrying only after attaining the age of 18 years and if possible, even later;

v)   to gain a better understanding of their environment related social issues and the impact on their lives; and

vi) to encourage adolescent girls to initiate various activities to be productive and useful members of the society.

   •   The Department considers that a single tailor-made Scheme for adolescent girls may not be able to achieve the objectives of Kishori Shakti Yojana as mentioned in para 8 above. There should be a basket of programmatic options available with the State/ UT/ districts to selectively intervene for the development of the adolescent girls on the basis of State/ UT/ area specific needs and requirements. Some of the options are provided below in para 10 and the State/ UT/ district may like to choose one or more of these, for empowerment of adolescent girls.

Child Development Schemes

1. Integrated Child Development Services (ICDS)

2. General Grant-in-aid (GIA) Scheme for Assistance to Voluntary Organisations in the field of Women and Child Development

1.   Integrated Child Development Services(ICDS)

The Integrated Child Development Service (ICDS) program was launched in the year 1975, in pursuance of the National Policy for Children. It is by far the only major national program that seeks to provide for children under the age of six.

Objectives:

- To improve the nutritional and health status of children below the age of 6 years.
- To lay the foundation for the proper psychological, physical and social development of the child.
- To reduce the incidence of mortality, morbidity, malnutrition and school dropouts.
- To achieve effective coordination of policy and implementation among various departments to promote child development.
- To enhance the capability of the mother to look after the normal health, nutritional and developmental needs of the child through proper community education.
- The nutrition component varies from state to state but usually consists of a hot meal cooked at the Anganwadi by the Anganwadi worker, based on a mix of pulses, cereals, oil, vegetable, sugar, iodized salt, etc. Nutrition and health education is imparted through counseling sessions, home visits and demonstrations. It covers issues such as infant feeding, family planning, sanitation and utilization of health services.
- The primary goal of ICDS is to break the inter-generational cycle of malnutrition, reduce morbidity and mortality caused by nutritional deficiencies by providing the following six services as a package through the network of Anganwadis.
  - Supplementary nutrition (SNP)
  - Non-formal pre-school education (PSE)
  - Immunisation
  - Health check-up
  - Referral services
  - Nutrition and Health Education (NHE)
  - The three services, viz. immunisation, health check-up and referral, are designed to be delivered through the primary health care infrastructure. While providing SNP, PSE and NHE are the primary tasks of the Anganwadi Centre, the responsibility of coordination with the health functionaries for provision of other services rests with the Anganwadi worker (AWW).
- ICDS is designed to provide services to children, pregnant women (PW), lactating mothers (LM) and adolescent girls (AG). While services to children are expected to yield results in the short run by contributing to reduction in child mortality and morbidity, those provided to PW are aimed at reducing the Maternal Mortality Rate (MMR) in the short run. The inclusion of LM is intended to address the high rate of Infant Mortality Rate (IMR), while the programmes for AGs address malnutrition with a long-term perspective. In this way, ICDS is expected to contribute to attainment of the following Millennium Development Goals (MDGs):

- Reduction in severe to moderate malnutrition among children (MDG-1)
- Reduction in IMR, CMR, MMR (MDG 4,5)
- Increase in enrollment, retention rates and reduction in dropout (MDG-2) by laying foundation at AWC.

1. General Grant-in-aid (GIA) Scheme for Assistance to Voluntary Organisations in the field of Women and Child Development

   The Ministry of Women and Child Development is implementing a Scheme called the General Grant-in-Aid Scheme, also called Scheme for innovative projects for women and children. The objective of the Scheme is to support innovative voluntary action and initiatives to render services for women and children. This Scheme is meant to supplement the existing schemes of the Ministry and of the Central Social Welfare Board and not to duplicate them. Financial assistance is provided for services that are not covered by the structured schemes of the Ministry or CSWB

2. Projects falling under the following categories may be supported under the Scheme:-
   a) Project that suggests a new approach to tackle a pressing social problem.
   b) Project which fills essential gaps in existing services and complements them so as to maximise the impact.
   c) Project to tackle a problem area which is relatively unserviced but where need is urgent.
   d) Project which provides integrated services, all the components need not be financially supported by one source.
   e) Project which is community based and renders non-institutional services. Where the nature of the problem so demands, institutional programmes will also be supported.

3. Pilot projects in any of the above areas will also be supported. Financial assistance will be provided to establish the innovative project over the project cycle, and will not be extended on a continuing basis as a matter of course.

4. Financial assistance is given up to 90% of the approved cost on recurring and non-recurring expenditure and the balance of 10% is to be met by the voluntary agency or any other organisation, but preferable by the voluntary organisation itself. In the case of an organisation located in remote, backward and tribal areas, the Government may bear 95% of the approved cost. The extent of financial assistance is decided on merit on a case to case basis by the Project Sanctioning committee under the Secretary, Ministry of Women and Child Development.

5. Interested organisations, with experience of at least 2 years, can apply through the State Govt/UT Admn. The application form and details of the Scheme may be downloaded from the website of the Ministry (www.wcd. nic.in).

References

1. Acker, J. (1990) Hierarchies, jobs, bodies: a theory of gendered organizations. Gender and Society 4 (2), pp. 39-158.

2. Adolph, B. (2003) The Role of Self-Help Groups in Rural Non-Farm Employment. Discussion paper prepared by the National Resources Institute, U.K. for DFID and the World Bank.

3. Advance Women's Health Kōkiritia Te Hauora Wāhine: A Checklist for Regional Health Authorities and Crown Health Enterprises, Ministry of Women's Affairs, Wellington, 1995

4. Agarwal, B. (1988) Neither sustenance nor sustainability: agricultural strategies, ecological degradation and Indian women in poverty, in B. Agarwal (ed.) Structures of Patriarchy: State, Community and Household in Modernising India. New Delhi: Kali for Women and London: Zed Books.

5. Agarwal, B. (1994) A Field of One's Own: Women and Land Rights in South Asia. Cambridge: Cambridge University Press.

6. Agarwal, B. (2001) Participatory exclusions, community forestry and gender: An analysis for South Asia and a conceptual framework. World Development 29 (10), pp. 1623-1648.

7. Agarwal, B. (2003) Gender and land rights revisited: Exploring new prospects via the state, family and market, in S. Razavi (ed.) Agrarian Change, Gender and Land Rights, Oxford: Blackwell Publishing (for the UN Research Institute for Social Development).

8. Ahmed, S. (1999) Changing gender roles in irrigation management: Sadguru's liftirrigation co-operatives. Economic and Political Weekly 34 (51).

9. Ahmed, S. (2002) Engendering organizational practice in NGOs: the case of UTTHAN. Development in Practice 12 (3-4), pp. 298-311.

10. Ahmed, S. (2002) Mainstreaming gender equity in water management: institutions, policy and practice in Gujarat, India, in Natural Resources Management and Gender: A Global Source Book. Amsterdam: Royal Tropical Institute (KIT) and Oxford: Oxfam.

11. All About Women in New Zealand, Statistics New Zealand, Wellington, 1993

12. Arun, S. (1999) Does land ownership make a difference? Women's roles in agriculture in Kerala, India. Gender and Development 7 (3), pp. 19-27.

13. Athreya, V.B. (2002) A note on agricultural labour: some gender issues. Teaching note prepared for Course on Gender Issues in Agriculture, Thrissur: Kerala Agricultural University.

14. Athreya,V.B. et al. (1990) Barriers Broken: Production Relations and Agrarian Change in Tamil Nadu. New Delhi: Sage.

15. Banerjee, N (1985) Women Workers in the Unorganised Sector: the Calcutta Experience. Sangam Books. India.

16. Banerjee, N. (1995) Grassroots Empowerment (1975-1990): A discussion paper. Occasional Paper No. 22, New Delhi: Centre for Women's Development Studies.

17. Batliwala S. (1993) Empowerment of Women in South Asia: Concepts and Practices. Sri Lanka: ASPBAE.

18. Baxi, Upendra (ed.) (1999) Reconstructing the Republic. Har-Anand, New Delhi.

19. BRIDGE Briefing Paper (2001) Feminisation of Poverty. Sussex: Institute of Development

20. Briefing to the Incoming Government He Whakamōhio Atu I te Kāwanatanga Hou, Ministry of Women's Affairs, Wellington, November 1993

21. Candida March, Ines Smyth, and Maitrayee Mukhopadhyay, (1999), a Guide to Gender-Analysis Frameworks, London: Oxfam Publishing

22. Carr, Marilyn et al. (1996) Speaking Out: Women's Economic Empowerment in South Asia. New Delhi: Vistaar Publications.

23. Chakrabarti, P.G.D. (2001) Gender Inequality and Social Policy in India. Paper presented at the International Sociological Association International Conference, Oviedo, Spain.

24. Chakravarti, Uma (2003) Gendering Caste: Through a Feminist Lens. Calcutta: Stree.

25. Chambers, R. (1988) Poverty in India: Concepts, Research and Reality. IDS Discussion Paper 241, Sussex: Institute of Development Studies.

26. Chambers, R. and G. Conway (1991) Sustainable Rural Livelihoods: Practical Concepts for the 2151 century. IDS Discussion Paper 296, Sussex: Institute of Development Studies. Chen, M. et al. (1986) Indian Women: A Study of their Role in the Dairy Movement. New Delhi: Shakti Books.

27. Chaturvedi S V.,(2015). Participatory Gender Auditing: As A Process For Gender Mainstreaming. Journal of Management Engineering and Information Technology, 2(1):2-6

28. Chen, M. and A. Dholakia (1986) SEWA's Women Dairy Co-operatives: A case study from Gujarat, in M. Chen et al. (eds.) Indian Women: A Study of their Role in the Dairy Movement. Ghaziabad: Vikas Publications.

29. Chen, M.A. (1991) Coping with Seasonality and Drought. New Delhi: Sage Publications.

30. Chen, M.A. (1993) Women and Wasteland Development in India: An Issue Paper, in A.M. Singh and N. Burra (eds.) Women and Wasteland Development in India. New Delhi: Sage.

31. Choudhury, A.R. (2004) Community institutions and gender: fishworkers in Kasargode, Kozhikode and Thiruvananthapuram, in S. Krishna (ed.) Livelihood and Gender, New Delhi: SAGE.

32. Christian Aid (2000) Selling suicide: farming, false promises and genetic engineering in developing countries.

33. CIDA (1993) Women and Fisheries Development. Hull: Canadian International Development Agency.

34. CIDA (1997) Guide to Gender Sensitive Indicators. Hull (Quebec).

35. Cooke, B. and U. Kothari (eds.) (2001) Participation: the new tyranny? London: Zed Books.

36. Counting for Nothing: What Men Value and What Women Are Worth, Marilyn Waring, Allen & Unwin: Port Nicholson Press, Wellington, 1988

37. Cross, N. (2002) Sustainable development explained. Developments 18, pp. 6-8.

38. Development Assistance Committee (DAC), 1998, 'Gender, Equality and Culture', in DAC Source Book on Concepts and Approaches linked to Gender Equality, OECD, Paris

39. DFID (1999) India: Country Strategy Paper. New Delhi / London (www.dfid. org).

40. DFID (2003) Poverty and Climate Change: Reducing the Vulnerability of the Poor through. Adaptation. London (www.dfid.org).

41. Doane, Donna L (1999) Indigenous Knowledge, Technology blending and gender implications. Gender and Technology Development 3 (2), May-Aug.

42. Dube, Leela (1997) Women and kinship: comparative perspectives on gender in south and South-East Asia. New Delhi: Vistaar.

43. Dube, Leela (2001) Anthropological Explorations in Gender. New Delhi: Sage Publications,

44. Duvvury, N. (1989) Women in Agriculture: A review of the Indian literature. Economic and Political Weekly Odober, pp. WS-96 to WS-112.

45. Duvvury, N. (1998)Women and agriculture in the new economic regime, in M. Krishnaraj et al. (eds.) Gender, Population and Development. New Delhi: OUP

46. EI-Bushra, J. (2000) Re-thinking gender and development practice for the twenty-first century. Gender and Development 8 (1), pp. 55-62.

47. Enarson, E. (1998) Through women's eyes: a gendered research agenda for disaster social science. Disasters 22(2), pp. 157-173.

48. Enarson, E. and B.H. Morrow (eds.) (1998) The Gendered Terrain of Disaster: Through Women's Eyes. Westport, CT: Greenwood Publications.

49. Families, Money and Policy: Summary of the Intra Family Income Study, Susan Kell Easting and Robin Fleming, Intra Family Income Study, Wellington, and Social Policy Research Centre, Massey University, Palmerston North, 1994

50. Fausto-Sterling, Anne (1992) Myths of Gender: Biological theories about women and men. New York: Basic Books.

51. Feldstein, H.5. and J. Jiggins (1994) Tools for the Field: Methodologies-Handbook for Gender Analysis in Agriculture. Connecticut: Kumarian Press.

52. Feldstein, H.S. and Poats, S.V 1989. Working Together: Gender Analysis in Agriculture 1: Case Studies Connecticut: Kumarian Press.

53. Feldstein, H.S. and Poats, S.v. 1989. Working Together: Gender Analysis in Agriculture 2: Case Studies. Connecticut: Kumarian Press.

54. Fernandez, A.P. (2001) Putting Institutions First-Even in Micro Finance: the MYRADA Experience. Bangalore: MYRADA.

55. Fernando, P. and V. Fernando (eds.) (1997) South Asian Women: Facing Disasters, Securing Life. Colombo: Intermediate Technology Publications for Duryog Nivaran.

56. Fordham, M. (1999) The intersection of gender and social class in disaster: balancing resilience and vulnerability. International Journal of Mass Emergencies and Disasters 17(1), pp. 15-36.

57. Geetha, V. (2002) Gender. Calcutta: Stree Publications.

58. Ghotge, N.5. and S.R. Ramdas (2002) Women and livestock: creating space and opportunities. LEISA India 4, (4), pp. 15-16.

59. Ghotge, N.S. (2004) Livestock and Livelihoods: The Indian Context. New Delhi: Foundation Books and Ahmedabad: Centre for E,nvironment Education.

60. Goetz, A.M. (1997) Managing organizational change: the gendered organization of space and time. Gender and Development 5 (1), pp. 17-27.

61. Goetz, A.M. and R. Sen Gupta (1996) Who takes the credit? Gender, power and control over loan use in rural credit programmes in Bangladesh. World Development 24(1), pp. 45-63.

62. Government of India (2002) Disaster Management: The Development Perspective. Tenth Five Year Plan 2002-2007, Chapter 7.

63. Groverman, V. and E. Walsum (1994) Women may lose or gain: expected impact of irrigation projects, in V. Gianotten (ed.) Assessing the Gender Impacts of Irrigation Projects. London: Intermediate Technology Development Group.

64. Groverman, V. and J.D. Gurung (2001) Gender and Organisational Change: Training ManuaL Kathmandu: International Centre for Integrated Mountain Development.

65. Guijit, I. and M.K. Shah (eds.) (1999) The Myth of Community: Gender Issues in Participatory Development. New Delhi: Vistaar Publications.

66. Hambly, H. et al. (2002) Gender and agriculture in the information society. ISNAR Briefing Paper # 55.

67. Harding Sandra (2000) Democratising Philosophy of Science for Local Knowledge movements: Issues and Challenges. Gender, Technology and Development 4 (1), Jan-April.

68. Harding, Sandra (1986) The Science Question in Feminism. Cornell University Press.

69. Harding, Sandra (ed.) (1987) Feminism and Methodology. Bloomington and Indiana: Indiana University Press.

70. Howard, Patricia (2003) The Major Importance of Minor Resources: Women and Plant Biodiversity. London: International Institute for Environment and Development.

71. Howard, Patricia (2003). Women and Plants: Gender Relations in Biodiversity Management and Conservation. London: ZED Books.

72. Howard-Borjas, P.L. (2001) Gender Relations in Local Plant Genetic Resource Management and Conservation. Netherlands: Wageningen University.

73. Hunt, Juliet 2000, 'Understanding gender equality in organisations: A tool for assessment and action', Development Bulletin, 51, 73–76.

74. ICSF (1997) Women First: Report of the Women in Fisheries Programme. ChennaL Indian Journal of Gender Studies 8 (2), July-Dec 2001. Special issue on Natural Resource Management

75. Kabeer, Naila 1994, Reversed Realities: Gender Hierarchies in Development Thought, Verso, London.

76. Labour Market 1994, Statistics New Zealand, Wellington, 1994

77. Locke, e. (1999) Constructing a gender policy for joint forest management in India. Development and Change 30 (2), pp. 265-285.

78. Longwe, Sara 1991, 'Gender awareness: The missing element in the Third World development project', in Tina Wallace and Candida March (eds), Changing Perceptions: Writings on Gender and Development, Oxfam, Oxford.

79. Macdonald, M., Springer, E. and 1.Dubel (1997) Gender and Organisational Change: Bridging the Gap between Policy and Practice. Amsterdam: Royal Tropical Institute.

80. Māori in Education: A Statistical Profile of the Position of Māori Across the New Zealand Education System, Davies and Nicholl, Ministry of Education, Wellington, 1993

81. March, Candida, Ines Smyth, and MaitraiyeeMukhopadhyay. (1999). A Guide to Gender-Analysis Frameworks. Oxford: Oxfam

82. March, e. et al. (1999) A Guide to Gender Analysis Frameworks. Oxford: Oxfam.

83. Mawdsley, E. (1998) After Chipko: from Environment to Region in Uttaranchal. Journal of Peasant Studies 25(4), pp. 36-54.

84. Mazumdar, V. (2000) Political Ideology of the Women's Movement's Engagement with Law. Occasional Paper No. 34, New Delhi: Centre for Women's Development Studies.

85. Mearns,. R (1999) Access to Land in Rural India: Policy Issues and Options. World Bank Working Paper.

86. Mehta, M. (1996) Our lives are no d.ifferent from that of our buffaloes: agricultural change and gendered spaces in a central Himalayan valley, in D. Rocheleau et al. (eds.) Feminist Political Ecology. London: Routledge.

87. Merrill-Sands, D. et al. (1999) Engendering organisational change: A case study of strengthening gender equity and organisational effectiveness in an international agricultural research institute, in A. Rao et al. (eds.) Gender at Work: Organisational Change for Equality. Connecticut: Kumarian Press.

88. MHHDC (2003) Human Development in South Asia 2002: Agriculture and Rural Development.

89. Mishra, P.K. et al. (2004) Gender and disasters: coping with drought and floods in Orissa, in S. Krishna (ed.) Livelihood and Gender. New Delhi: Sage.

90. Moench, M. and A. Dixit (eds.) (2004) Adaptive Capacity and Livelihood Resilience: Adaptive Strategies for Responding to Floods and Drought in South Asia. Colorado/ Kathmandu: Institute for Social and Environmental Transition.

91. Mohan Rao, J. (1998) Agricultural development under state planning, in T.J. Byres (ed.) The State, Development Planning and Liberalisation in India. New Delhi: Oxford University Press.

92. Mohanty, C. (1991) 'Under Western Eyes. Feminist Scholarship and Colonial Discourse' in Mohanty, C., Russo, A. and L. Torres (eds.), 1991, Third World Women and the Politics of Feminism, Bloomington, Indiana University Press

93. Moore, H. 1994, A Passion for Difference, Cambridge, Polity

94. Morrow K. (2002) Information villages: Connecting rural communities in India. LEISA India June, pp. 17-19.

95. Moser, C. (1993) Gender Planning and Development: Theory, Practice, and Training. Routledge: London, UK.

96. Moser, Caroline O 1989, 'Gender planning in the Third World: Meeting practical and strategic gender needs', World Development, 17(11), 1799–1825.

97. Moser, Caroline O 1993, Gender Planning and Development: Theory, Practice and Training, Routledge, London and New York.

98. MSSRF (2001) Food Insecurity Atlas of Rural India. Chennai.

99. MSSRF (2001) Like Paddy in Rock: Local Institutions and Gender Roles in Kolli Hills. Chennai.

100.MSSRF (2003) Farmers' Rights and Biodiversity: A Gender and Community Perspective. Chennai.

101.Mukhopadhyay, M. (2001) Introduction: women and property, women as property, in Gender Perspectives on Property and Inheritance: A Global Sourcebook. Amsterdam: KIT (Royal Tropical Institute) and Oxford: Oxfam.

102.Murthy, R. K. and N. Rao (1997) Addressing Poverty: Indian NGOs and their Capacity Enhancement in the 1990s. New Delhi: Friedrich Ebert Stiftung.

103.Murthy, R.K. (ed.) (2001) Building Women's Capacities: Interventions in Gender Transformation. New Delhi: Sage and ICCO.

104.New Zealand Now: Māori, Statistics New Zealand, Wellington, 1994

105.Njoki, Wane (2001) Narratives of Embu Rural women-gender roles and indigenous knowledge. Gender, Technology and Development Sep-Dec, 5 (3).

106.No. 55, The Hague: International Service for National Agricultural Research. (www.isnar.cgiar. org)

107.Nussbaum, M., and Glover, J., 1995, Women, Culture and Development: A Study of Human Capabilities, Clarendon Press, Oxford

108.Odame, H.H. et al. (2002) Gender and Agriculture in the Information Society. Briefing Paper

109.Omvedt, G. and G. Kelkar (1995) Gender and Technology: Emerging visions from Asia, Bangkok: Asian Institute of Technology and Chennai: IWID.

110.Overholt, Catherine, Kathleen Cloud, Mary Anderson and James Austin 1985, 'Women in development: A framework for project analysis' in Overholt, Catherine, Kathleen Cloud, Mary Anderson and James Austin, Gender Roles in Development Projects: A Case Book, Kumarian Press, West Hartford, Connecticut.

111.Oxaal, Z. and S. Baden (1997) Gender and Empowerment: Definitions, Approaches and Implications for Policy. BRIDGE Report No. 40, Sussex: Institute of Development Stuidies.

112.Oxfam, 1995, 'Women and Culture,' Gender and Development, Oxfam Journal, Vol.3, No.1, February, Oxfam, Oxford

113.Oxford: OUP/Mahbub ul Haq Human Development Centre. Mies, Maria and Vandana Shiva (1993) Ecofeminism. New Delhi: Kali for Women.

114.Pandey, Gopa (2002) Analysis of Gender Perception in Forest Management in India, in A decade of Joint forest Management-Retrospection and Introspection

115.Parasuraman, S. and P.v. Unnikrishnan (2000) India Disasters Report: Towards a Policy Initiative. New Delhi: Oxford University Press.

116.Parker, A Rani 1993, Another Point of View: A Manual on Gender Analysis Training for Grassroots Workers, UNIFEM, New York.

117.Platform for Action, Fourth World Conference on Women, Beijing, 1995

118.Prabhu, M. (1999) Marketing treadle pumps to women farmers in India. Gender and Development 7 (2), pp. 25-33.

119.Programme of Action, International Conference on Population and Development, Cairo, 1994

120.Quisumbing, A.R. et al. (1998) The importance of gender issues for environmentally and socially sustainable rural development, in E. Lutz (ed.) Agriculture and the Environment: Perspectives on Sustainable Rural Development. Washington, DC: World Bank.

121.Ramaswamy, U. et al (2000) Reconstructing Gender towards Collaboration. Bangalore: Books for Change.

122. Ramaswamy, Uma and Bhanumathy Vasudevan (1996) Gender in livestock sector in Andhra Pradesh: workbook. Working paper, Andhra Pradesh: Indo Swiss Project. p. 107.

123. Ramdas, S.R. et al. (2004) Overcoming gender barriers: local knowledge systems and animal health healing in Andhra Pradesh and Maharashtra, in S. Krishna (ed.) Livelihood and Gender: Equity in Community Resource Management. New Delhi: Sage.

124. Rao, A. and D. Kelleher (2003) Institutions, organizations and gender equality in an era of globalization. Gender and Development 11(1), pp. 142-149.

125. Rao, A., S. Stuart, and D. Kelleher (1999) Gender at Work: Organisational Change for Equality. West Hartford: Kumarian Press.

126. Rao, B. (1991) Women and water in rural Maharashtra. Environment & Urbanization (3) 2/ pp 57-65.

127. Sainath, P. (2004) The Agrarian Crisis-Some Gender Dimensions. Keynote Address at the Conference on Gender Concerns and Food Security Issues in Rice Livelihood Systems in India: Challenges and Opportunities. <;:hennai: MSSRF.

128. Sarin, M. (2002) Not Seeing the Wood for the Trees: Rural Women and Forests in a Globalising Context, in Women's Empowerment Policy and Natural Resources – What Progress? Bangalore: Write-Arm with DFID / UEA-ODG and the Planning Commission.

129. Sarin, M. et al. (1998) Who is Gaining? Who is Losing? Gender and Equity Concerns in Joint Forest Management. New Delhi: Society for the Promotion of Wasteland Development.

130. Satheesh, P.v. (1997) Genes, gender and biodiversity: Deccan Development Society's Community Genebanks, in L. Sperling and M. Loevinsohn (eds.) Using Diversity: Enhancing and Maintaining Genetic Resources on Farms. Ottawa: International Development Research Centre.

131. Saxena, N.C. (1993) Women and Wasteland Development in India: Policy Issues, in A.M. Singh and N. Burra (eds.) Women and Wasteland Development in India. New Delhi: Sage.

132. Sayers, J. (1982) Biological Politics: Feminist and Anti-Feminist Perspectives. London and New York: Tavistock.

133. Sen, A. (2001) Many faces of gender inequality. Frontline, Nov. 9, pp. 4-14.

134. Senthilkumaran, S. and S. Arunachalam (2002) Expanding the Village Knowledge Centres in Pondicherry. Regional Development Dialogue 23 (2).

135. Sharma, D (2002) Digital Library on Indian Medicine Systems: Another tool for Biopiracy in Economic and Political Weekly June 22

136. Sharma, K. (1998) Transformative politics: dimensions of women's participation in Partchayati Raj, Indian Journal of Gender Studies 5 (1), pp. 25-47.

137. Shiva, V. (1997) Biopiracy: The Plunder of Nature and Knowledge. Boston: South End Press.

138.Shiva, Vandana (1991) Most Farmers in India are Women. New Delhi: FAO.

139.Shiva, Vandana, A H Jafri and G Bedi (1997) The Enclosure and Recovery of the Commons: Biodiversity, Indigenous Knowledge and Intellectual Property Research. Foundation for Science, Technology and Ecology, New Delhi.

140.Shylendra, H. S. (1999) Micro-Finance and Self Help Groups (SHGs): A Study of the Experience of Two Leading NGOs, SEWA and AKRSP(I) in Gujarat, India. Research Paper No. 16, Anand: Institute of Rural Management, Anand (IRMA).

141.Singh, A.M. and N. Burra (eds.) (1993) Women and Wasteland Development in India. New Delhi: Sage.

142.Sinha, F. (1996) Women or We-men, who is in control? An assessment of the functioning of women's dairy co-operative societies, Report of a study conducted by EDA Rural Systems (Delhi) for the National Dairy Development Board.

143.Srinivasan, V. (1993) Indian Women-A Study of their Role in the Handicrgfts and Dairying Sectors. New Delhi: Har-Anand Publications.

144.Status of New Zealand Women, 1992: Second Periodic Report on the Convention on the Elimination of All Forms of Discrimination Against Women, Ministry of Women's Affairs, Wellington, 1992

145.Stephen, F. (2001) Empowering women in gram panchayats through training, in R.K. Murthy (ed.) Building Women's Capacities. New Delhi: Sage.

146.Studies.

147.Swaminathan, M. S (ed.) (1998) Gender Dimensions in Biodiversity Management. New Delhi: Konark Publishers.

148.Swatninathan, P. (1999) The Gendered Politics of Fuel in India. Gender, Technology and Development 3 (2), pp 165-187.

149.Sweetman, C. (1997) Editorial: gender and organizational change. Gender and Development 5(1), pp. 2-9.

150.Thamizoli, P. and the MSSRF team (2004) Mainstreaming gender concerns in mangrove conservation and management: the Pichavaram Coast, Tamil Nadu, in S. Krishna (ed.) Livelihood and Gender. New Delhi: Sage.

151.The Status of Girls and Women in New Zealand Education and Training, Fiona Sturrock, Ministry of Education, Wellington, 1993

152.UNDP (1997) Human Development Report. New Delhi, Oxford: Oxford University Press.

153.UNDP (2003) Human Development Report: The Millennium Development Goals-A compact among nations to end human poverty. New Delhi, Oxford: Oxford University Press.

154.USAID. 2012. Gender Equality and Female Empowerment Policy. Washington, D.C.:

155.Van Koppen, B. (2001) Gender in integrated water management: an analysis of variation. Natural Resources Forum 25, pp. 299-312.

156. Vasavada, S. (2000) Women irrigators and participatory irrigation management: policy and approaches to mainstream gender concerns-lessons from the Aga Khan Rural Support Programme, India. Paper presented at CIDA-Shastri Workshop on Empowering Rural Women? Policies, Institutions and Gendered Outcomes in Natural Resources Management, Anand: Institute of Rural Management, September.

157. Venkateswaran, S (1997) Environment, Development and the Gender Gap. New Delhi: Sage Publications.

158. Vishwanathan, M (1997) Women in Agriculture & Rural Development. Jaipur: Printwell.

159. Waldie, K and S. Ramkumar (2002) Improving Animal Husbandry Services for the Poor Women through Capacity Building in Gender Awareness. UK: DFID

160. Warren, Carol A. (1998) Gender Issues in Field; Researc,h. New Delhi: Sage Publications.

161. Whitehead, A. (1985) Effects of Technological Change on Rural Women: A Review of Analysis and Concepts, in I. Ahmed (ed.) Technology and Rural. Women.

162. Williams, S., J. Seed and A. Mau (1994) Oxfam Gender Training Module. UK and Ireland: Oxfam.

163. Wilson, K. (2002) The new micro-finance: An essay on the SHG movement in India. Journal of Micro-Finance 4 (2), pp. 217-245.

164. Women and Economics: A New Zealand Feminist Perspective, Prue Hyman, Bridget William Books, Wellington, 1994

165. World Bank, Food and Agriculture Organization, and International Fund for Agricultural Development. 2009. Gender in agriculture sourcebook. Washington, D.C.: USAID. http://worldbank.org/genderinag

166. Woroniuk, Beth, Johanna Schalkwyk and Helen Thomas 1997, Overview: Gender Equality and Emergency Assistance/Conflict Resolution, Division for Humanitarian Assistance, Sida, Stockholm.

167. Zwarteveen, M. (1997) Water: from basic need to commodity-a discussion on gender and water rights in the context of irrigation. World Development 25 (8).

168. Zwarteveen, M. and N. Neupane (1996) Free-riders or victims: women's non-participation in irrigation management in Nepal's Chhattis Mauja Irrigation Scheme. Research Report No. 7, Colombo: International Water Management Institute.

169. Other web sources and articles:

170. www.bigpond.com.kh/users/gad/glossary/gender.htm

171. global.finland.fi/julkaisut/taustat/nav_gender/glossary.htm

172. www.un-instraw.org

173. www.thecommonwealth.org

174. www.opsc.qld.gov.au

175. www.deir.qld.gov.au

176. www.un.org
177. www.adcq.qld.gov.au
178. www.eowa.gov.au
179. www.hreoc.gov.au
180. www.swccfc.gc.ca/pubs
181. info@women.qld.gov.au
182. www.women.qld.gov.au
183. www.unfpa.org
184. www.oesr.qld.gov.au
185. www.officeforwomen.sa.gov.au
186. www.community.wa.gov.au
187. www.swc-cfc.gc.ca
188. www.mwa.govt.nz
189. www.gdrc.org/gender/framework/framework.html
190. www.cge.org.za
191. http://www.dol.gov/oasam/regs/statutes/titleIX.htm
192. http://www.grcc.edu/womensstudies.
193. info@medinstgenderstudies.org.
194. www.glbtq.com
195. http://www.ilo.org/public/libdoc/ILOThesaurus/ english/tr35.htm
196. http://arabstates.undp.org/contents/file/GenderMainstreamingTraining.pdf
197. http://arabstates.undp.org/contents/file/GenderMainstreamingTraining.pdf
198. http://ipsnews.net
199. http://hdr.undp.org/
200. http://www.siyanda.org/docs_gem/index_implementation/genderman.htm
201. http://www.who.int/reproductivehealth/gender/glossary.html
202. http://www.iom.int/jahia/jsp/index.jsp
203. http://ezinearticles.com/?Definition-of-Feminism&id=1697184
204. http://depts.drew.edu/wmst/CoreCourses/WMST112/WMST112_Glossary.htm
205. http://www.undp.org
206. http://www.siyanda.org/docs_gem/index_implementation/genderman.htm
207. http://www.bridge.ids.ac.uk/reports.html
208. http://www.ifad.org/gender/glossary.htm
209. http://icds-wcd.nic.in/gbhb/Link%20hand%20pdf/Hand%20Book%20Chap%202.pdf
210. http://drd.nic.in/drd/downloads/programmes-schemes/Gender_Audit.pdf
211. www.indiastat.com
212. www.agricoop.nic.in

213. www.vistar.nic.in
214. www.vikaspedia.in
215. http://eige.europa.eu/gender-mainstreaming/what-is-gender-mainstreaming
216. www.ndi.org/files/Guide%20to%20Gender%20Analysis%20Frameworks.pdf
217. http://www.snvworld.org/gender/gender-mainstreaming_analysis_1.htm,
218. International Labour Organization, "Online Gender Learning and Information Module."
219. Netherlands Development Organization, "Gender Reference Guide."

www.ingramcontent.com/pod-product-compliance
Lightning Source LLC
Chambersburg PA
CBHW031950180326
41458CB00006B/1681